Science and Fiction

Science and Fiction – A Springer Series

This collection of entertaining and thought-provoking books will appeal equally to science buffs, scientists and science-fiction fans. It was born out of the recognition that scientific discovery and the creation of plausible fictional scenarios are often two sides of the same coin. Each relies on an understanding of the way the world works, coupled with the imaginative ability to invent new or alternative explanations—and even other worlds. Authored by practicing scientists as well as writers of hard science fiction, these books explore and exploit the borderlands between accepted science and its fictional counterpart. Uncovering mutual influences, promoting fruitful interaction, narrating and analyzing fictional scenarios, together they serve as a reaction vessel for inspired new ideas in science, technology, and beyond.

Whether fiction, fact, or forever undecidable: the Springer Series "Science and Fiction" intends to go where no one has gone before!

Its largely non-technical books take several different approaches. Journey with their authors as they

- Indulge in science speculation—describing intriguing, plausible yet unproven ideas;
- Exploit science fiction for educational purposes and as a means of promoting critical thinking;
- Explore the interplay of science and science fiction—throughout the history of the genre and looking ahead;
- Delve into related topics including, but not limited to: science as a creative process, the limits of science, interplay of literature and knowledge;
- Tell fictional short stories built around well-defined scientific ideas, with a supplement summarizing the science underlying the plot.

Readers can look forward to a broad range of topics, as intriguing as they are important. Here just a few by way of illustration:

- Time travel, superluminal travel, wormholes, teleportation
- Extraterrestrial intelligence and alien civilizations
- Artificial intelligence, planetary brains, the universe as a computer, simulated worlds
- Non-anthropocentric viewpoints
- Synthetic biology, genetic engineering, developing nanotechnologies
- Eco/infrastructure/meteorite-impact disaster scenarios
- Future scenarios, transhumanism, posthumanism, intelligence explosion
- Virtual worlds, cyberspace dramas
- Consciousness and mind manipulation

More information about this series at http://www.springer.com/series/11657

Andrew May

Rockets and Ray Guns: The Sci-Fi Science of the Cold War

 Springer

Andrew May
Crewkerne, United Kingdom

ISSN 2197-1188   ISSN 2197-1196   (electronic)
Science and Fiction
ISBN 978-3-319-89829-2   ISBN 978-3-319-89830-8   (eBook)
https://doi.org/10.1007/978-3-319-89830-8

Library of Congress Control Number: 2018941553

Printed on acid-free paper

This Springer imprint is published by the registered company Springer International Publishing AG part of Springer Nature.
The registered company address is: Gewerbestrasse 11, 6330 Cham, Switzerland

Contents

The Super-Bomb

In which scientists discover huge amounts of energy locked up inside the atom, but insist it could never be liberated on a significant scale. Science fiction authors, on the other hand, became fascinated with the idea of atomic power and its potential to create awesome new super-weapons. Eventually the rest of the world caught up, and science fiction became science fact when the atom bombs fell on Hiroshima and Nagasaki. But instead of ushering in a new age of devastatingly destructive warfare, the sheer power of nuclear weapons led to an uneasy kind of peace—the Cold War—and even that had been prophetically anticipated by authors like George Orwell and Arthur C. Clarke.

Secrets of the Atom

Many scientific terms are based on ancient Greek words, but very few of them actually originated in ancient Greece. "Atom" is one of the few exceptions. It was first used, in close to its modern sense, by the philosopher Democritus in the fifth century BCE.

Here is what Isaac Asimov—best known for his science fiction, but also a prolific writer on science fact—had to say about Democritus in his *Biographical Encyclopedia of Science and Technology*:

> He is best known for his atomic theory. He believed that all matter consisted of tiny particles, almost infinitesimally small, so small that nothing smaller was conceivable. Hence they were indivisible; the very word "atom" means "indivisible".…. The

© Springer International Publishing AG, part of Springer Nature 2018
A. May, *Rockets and Ray Guns: The Sci-Fi Science of the Cold War*, Science and Fiction,
https://doi.org/10.1007/978-3-319-89830-8_1

atoms, said Democritus, differed from each other physically, and in this difference was to be found an explanation for the properties of various substances. . ... Apparent changes in the nature of substances consisted merely in the separation of joined atoms and their rejoining in a new pattern. [1]

As a philosopher rather than a scientist, Democritus didn't look for any physical evidence to support his view, so it remained a matter of opinion. It wasn't a widely shared opinion, either, and atomic theory was largely ignored for more than two millennia. It only really came to the fore in the 19th century, by which time much more was known about physics and chemistry. It occurred to the English chemist John Dalton that atoms could be used to explain the otherwise baffling differences between, say, carbon dioxide and carbon monoxide. Here is Asimov again:

It seemed to Dalton that carbon monoxide might be composed of one particle of carbon united with one particle of oxygen (where the oxygen particle was four-thirds as heavy as the carbon particle) while carbon dioxide was composed of a particle of carbon combined with two oxygen particles. . ... Dalton recognized the similarity of this theory to that advanced by Democritus 22 centuries earlier and therefore called these tiny particles by Democritus's own term, "atoms". [2]

Although Dalton's theory had great explanatory power, scientists were still reluctant to accept atoms as real phenomena rather than philosophical abstractions. To quote science writer Brian Clegg:

As late as the early 20th century, there was still doubt as to whether atoms really existed. In the early days of atomic theory, atoms were considered by most to be useful concepts that made it possible to predict the behaviour of materials without there being any true, individual particles. [3]

But the journey had only just begun. In Dalton's theory, the only difference between atoms of different elements lay in their mass. Hydrogen was the lightest known atom, while the heaviest—of the naturally occurring elements known in the 19th century—was uranium. It was the latter which proved to be the key that transformed atomic science from a neat theory into something of enormous practical importance. In the words of journalist Piers Bizony:

Uranium's more dramatic potentials emerged in 1896, when Becquerel discovered quite by accident that it gave off invisible rays capable of fogging photographic plates wrapped in light-proof black paper. [4]

What the French physicist Henri Becquerel had stumbled upon was the phenomenon of radioactivity. It meant that Dalton's atoms were no longer just an abstract way of describing chemical reactions. Under certain circumstances, they could produce energy in a form that was previously completely unknown.

Becquerel's discovery, at the very end of the 19th century, triggered a revolution in atomic physics that went on to dominate the first three decades of the 20th century. The driving force behind much of this research was a New Zealand born scientist named Ernest Rutherford. Less well known than his near-contemporary, Albert Einstein—who we will come to in due course—Rutherford's work was arguably just as important.

Rutherford was a pioneer of modern science in more ways than one. At the start of his career, scientific research was largely in the hands of individuals and small teams. But Rutherford—particularly after he became director of Cambridge University's Cavendish Laboratory in 1919—led the trend away from this, towards the larger, multi-national collaborative projects that we associate with "Big Science" today.

Together with his co-workers, Rutherford unravelled the substructure of the atom, and showed how it was responsible for the radiation that Becquerel had discovered. Rutherford's work also opened up the possibility of "nuclear reactions"—analogous to everyday chemical reactions, but involving much higher energies.

Rutherford pictured the atom as consisting of a dense, positively-charged nucleus—that's where the word "nuclear" comes from—surrounded by smaller, negatively-charged electrons. He originally envisaged these orbiting around the nucleus like planets around the Sun, but that picture was modified by one of Rutherford's colleagues, the Danish physicist Niels Bohr. For consistency with quantum theory, Bohr showed the electrons must be confined to discrete energy shells.

As to the nucleus itself—by 1920, Rutherford was convinced this was made up of smaller particles, in the form of positively charged protons and uncharged neutrons. This picture was confirmed in 1932, when his student James Chadwick found experimental evidence for the neutron. In its final form, the "Rutherford-Bohr model" is still considered a good approximation to the true structure of the atom today (see Fig. 1).

A particular atomic nucleus can be described by just two figures: the number of protons and the number of neutrons it contains. The number of protons (which is equal to the number of surrounding electrons) is called the "atomic number", while the total number of protons and neutrons together is called the "atomic weight". The chemical properties of an element are determined by its electrons, and so depend only on its atomic number. However, an

4   A. May

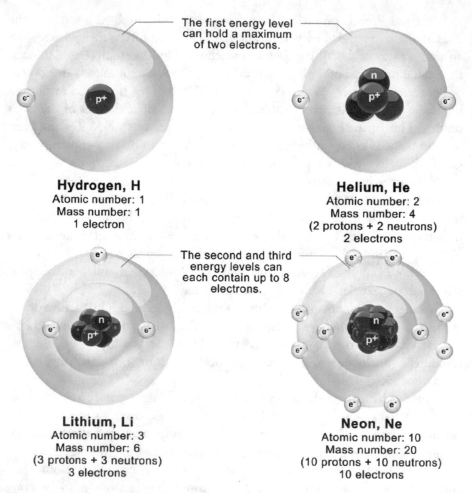

The first energy level can hold a maximum of two electrons.

The second and third energy levels can each contain up to 8 electrons.

Hydrogen, H
Atomic number: 1
Mass number: 1
1 electron

Helium, He
Atomic number: 2
Mass number: 4
(2 protons + 2 neutrons)
2 electrons

Lithium, Li
Atomic number: 3
Mass number: 6
(3 protons + 3 neutrons)
3 electrons

Neon, Ne
Atomic number: 10
Mass number: 20
(10 protons + 10 neutrons)
10 electrons

Fig. 1 The Rutherford-Bohr model of the atom, in the case of four different elements: hydrogen, helium, lithium and neon. The small central nuclei are made up of protons (blue) and neutrons (red), surrounded by a cloud of electrons (Wikimedia user BruceBlaus, CC-BY-3.0)

element of given atomic number may have a varying number of neutrons, and hence varying atomic weight. Different forms of an element with different atomic weights are referred to as "isotopes"—and while their chemical properties are identical, their nuclear properties may vary considerably.

In particular, some isotopes are stable while others are unstable. In effect, an unstable nucleus has more energy than it needs. As a result, it has a natural tendency to change into a more stable form, releasing its excess energy in the process. In the case of uranium, for example, the most stable isotope has an atomic weight of 238, while the commonest of its unstable isotopes, U-235, has three fewer neutrons. Paradoxically, that shortage of neutrons translates to an excess of energy—and U-235 only needs a small push to make it break up into more stable nuclei and release that excess energy in the form of radiation.

There's an analogy here with the well-established chemistry of high explosives. In simple terms, these consist of complex, unstable molecules which have a natural inclination to break down into simpler, more stable ones. When they do this, they release excess "chemical energy"—with explosive results.

Looked at more closely, a molecule's chemical energy is really just the electrical energy that holds it together. This is one instance of the more general concept of potential energy: hidden energy that can be converted into more active forms under the right circumstances. This immediately begs the question—is there an analogous kind of potential energy lurking inside an unstable nucleus like U-235?

The existence of natural radioactivity—apparently energy from nowhere—shows that there must be. Ultimately it comes from the binding energy inside the nucleus itself—the force that holds the protons and neutrons together. Unlike molecular binding energy, it's not electrical in nature—in fact it's much more powerful. And just like the excess binding energy in a chemical high explosive, it could—in principle—be made into a weapon.

As obvious as this may seem in hindsight, scientists were slow to pick up on it. Science fiction writers got there much quicker.

Potential Energy

Writing about the prodigious speed with which SF writers recognized the destructive potential of atomic energy, genre historian Peter Nicholls said:

> One almost forgotten writer, Robert Cromie got there through some extraordinary leap of the imagination even before the discovery of radioactivity by Henri Becquerel in 1896. In 1895, Cromie wrote of the power locked in the atom in his novel *The Crack of Doom*, in which a mad scientist utilizes this principle to build a bomb, and holds the world to ransom. Atom bombs developed a rapid popularity in science fiction after Rutherford's work. George Griffith wrote of atomic missiles in *The Lord of Labour* (1911), and H. G. Wells envisaged the effects of the atom bomb in *The World Set Free* (1914). [5]

The last-named work deserves a closer look. Wells wrote *The World Set Free* in 1914, just a few months before the outbreak of World War One. The novel was informed by Wells's reading on the then-topical subject of radioactivity—with, for example, Rutherford being mentioned by name in the first line of the first chapter. Even before this, in a prelude set in Wells's own time, a professor makes the following remarks in a public lecture:

We know now that the atom, that once we thought hard and impenetrable, and indivisible and final and lifeless, is really a reservoir of immense energy. . . . This little bottle contains about a pint of uranium oxide; that is to say, about 14 ounces of the element uranium. It is worth about a pound. And in this bottle, ladies and gentlemen, in the atoms in this bottle there slumbers at least as much energy as we could get by burning 160 tons of coal. If at a word, in one instant I could suddenly release that energy here and now it would blow us and everything about us to fragments. [6]

In Wells's future history, it takes the world several decades to solve that problem—and not initially in a military context:

It was in 1953 that the first Holsten-Roberts engine brought induced radioactivity into the sphere of industrial production, and its first general use was to replace the steam engine in electrical generating stations. [6]

This, for someone writing in 1914, is an extraordinarily good guess. In the real world, the first use of a nuclear reactor for electrical power generation occurred in the Soviet Union in 1954—so Wells was only out by a year. He was, however, less accurate in placing the first atomic bombs—not in 1945, as actually happened, but in a fictional war of 1956. His bombs, moreover, bear little resemblance to their real-world counterparts—but they're just as fearsome:

A moment or so after its explosion began it was still mainly an inert sphere exploding superficially, a big, inanimate nucleus wrapped in flame and thunder. Those that were thrown from aeroplanes fell in this state, they reached the ground still mainly solid, and, melting soil and rock in their progress, bored into the Earth. There . . . the bomb spread itself out into a monstrous cavern of fiery energy at the base of what became very speedily a miniature active volcano . . . spinning furiously and maintaining an eruption that lasted for years or months or weeks according to the size of the bomb employed. [6]

In other words, Wells pictured an atomic explosion as differing from a conventional one in duration rather than magnitude. Although that's not how things actually worked out, it does foreshadow the long-term devastation of radioactivity that became the most fearsome hallmark of nuclear weapons. Here is Wells's description of an "after the bomb" Paris:

> Few who adventured into these areas of destruction and survived attempted any repetition of their experiences. There are stories of puffs of luminous, radioactive vapour drifting sometimes scores of miles from the bomb centre and killing and scorching all they overtook. And the first conflagrations from the Paris centre spread westward half-way to the sea. Moreover, the air in this infernal inner circle of red-lit ruins had a peculiar dryness and a blistering quality, so that it set up a soreness of the skin and lungs that was very difficult to heal. [6]

As in so many other things, Wells was ahead of his time. It was not long, however, before other writers caught up—and not just in the science fiction genre. For example, Agatha Christie's mystery novel, *The Big Four* (1927), pits her famous detective Hercule Poirot against an archetypal "mad scientist" named Madame Olivier. As Poirot explains to the novel's narrator, Captain Hastings:

> Madame Olivier's experiments have proceeded further than she has ever given out. I believe that she has, to a certain extent, succeeded in liberating atomic energy and harnessing it to her purpose. [7]

These words are borne out a few pages later, as Hastings and Poirot experience an atomic explosion for themselves:

> The Earth shook and trembled under our feet, there was a terrific roar and the whole mountain seemed to dissolve. We were flung headlong through the air. [8]

The real home of atomic speculation, however, was in the SF pulp magazines that began to appear in the late 1920s. In spite of their lowbrow image, many of the stories they printed contained a fair amount of "real science"—something pioneering editor Hugo Gernsback was at pains to emphasize. Introducing a story called "When the Atoms Failed" in the January 1930 issue of *Amazing Stories*, he proudly ascribed it to "our new author, who is a student at the Massachusetts Institute of Technology" [9].

The author in question was John W. Campbell—who would go on to become the editor of his own magazine, *Astounding Science Fiction*, a few years later. At the time, however, he really was just a student—and "When the Atoms Failed" was his first foray into SF. Like *The World Set Free* and *The Big*

Fig. 2 A dramatic scene from "When the Atoms Failed" by John W. Campbell, from *Amazing Stories*, January 1930 (public domain image)

Four it features devastating atomic weapons—not bombs in this case, but a kind of death-ray (see Fig. 2).

In retrospect, the most remarkable thing about Campbell's story is the way he quantifies his fictional technology in terms of real scientific principles, rather than just throwing buzzwords around. The protagonist constructs a super-computer—in itself a sophisticated idea for 1930—to help evaluate the complex mathematical equations associated with nuclear physics. As he puts it:

> I developed that machine further in my laboratory, and carried it far beyond the original plans. I can do with it a type of mathematics that was never before possible, and that mathematics, on that machine, has done something no man ever did. . . .. It had reached the ultimate, definitive equation of all matter! This final equation gave explicit instructions to the understanding; it told just how to completely destroy matter. It told how to release such terrific energy, I was afraid

to try it. . . . The energy of matter has been known for many years; simple arithmetic can calculate the energy in one gram of matter. One gram is the equivalent of about ten drops of water and that much matter contains 900,000,000,000,000,000,000 ergs of energy, all this in ten drops of water! . . . Material energy is 10,000,000,000 times as great as the energy of coal. Perhaps now you can see why I was afraid to try out those equations. One gram of matter could explode as violently as 7,000 tons of dynamite! [9]

When Campbell says "the energy of matter has been known for many years" he's referring to the famous equation $E = mc^2$, first proposed by Albert Einstein in 1905. This defines the total energy E associated with a mass m of matter—the constant of proportionality being the square of the speed of light, c (300,000,000 m/second). Campbell uses an old unit, the "erg", to measure energy; nowadays we use joules (1 joule = 10,000,000 ergs). In modern units, the energy E associated with a kilogram of matter is 90,000,000,000,000,000 joules—roughly equivalent to a gigawatt of power sustained over three years.

The idea that any perfectly ordinary—and apparently inert—object contains such an enormous amount of energy was as fascinating to SF writers as it was bewildering to the general public. As Einstein himself wrote much later:

If every gram of material contains this tremendous amount of energy, why did it go so long unnoticed? The answer is simple enough: so long as none of the energy is given off externally, it cannot be observed. [10]

This is where SF diverges from reality. While Campbell's super-computer "told just how to completely destroy matter", that knowledge didn't exist in the real world at the time the story was written.[1] More than that—the very idea that it might exist wasn't entertained by serious scientists. Einstein's biographer, Walter Isaacson, recounts an encounter the great man had in 1919 which makes that perfectly clear:

A young man . . . insisted on showing him a manuscript. On the basis of his $E = mc^2$ equation, the man insisted, it would be possible "to use the energy contained within the atom for the production of frightening explosives". Einstein brushed away the discussion, calling the concept foolish. [11]

---

[1] It's now known that the complete annihilation of matter—and the consequent release of all that energy—can indeed be achieved, but only in situations where matter comes into direct contact with antimatter. That's not what happens in an atom bomb, where only a small fraction of the total mass is converted to explosive energy.

The first step toward proving the anonymous young man right—and Einstein wrong—came just over a decade later, in 1932. Rutherford's team at the Cavendish laboratory in Cambridge succeeded in splitting lithium nuclei into two smaller pieces, called alpha particles, by bombarding them with fast-moving protons. As Piers Bizony explains:

> The Cavendish team worked out that the disintegration of the lithium nucleus into two alpha particles accounted for almost all the original mass. But not quite. Two per cent of the mass had vanished. This, they realized, had been converted into the energy required to throw those alpha particles out of the nucleus with such staggering force. It was as if an unimaginably powerful pent-up spring of energy had been released. It was the first practical demonstration that Albert Einstein's famous equation, $E = mc^2$, was correct. Matter was indeed an incredibly compacted form of energy, and the compression factor was the square of the speed of light. An infinitesimally small amount of matter could be persuaded to release a vast amount of energy. [12]

The device used in these experiments was a particle accelerator. Colloquially known as an "atom smasher", this was the ancestor of huge machines like the Large Hadron Collider today. Such machines deal with energies that are enormous on the microscopic scale of atoms—but still extremely tiny when viewed in a macroscopic, real world context. This situation led to one of Rutherford's most notorious pronouncements, in a 1933 interview with *The New York Herald Tribune*:

> The energy produced by the breaking down of the atom is a very poor kind of thing. Anyone who expects a source of power from the transformation of these atoms is talking moonshine. [13]

As short-sighted as this statement looks in hindsight, it was perfectly true in terms of the known physics of the time. On top of that, Rutherford's scepticism would have been boosted by the fact that "atomic power" had become a fashionable buzzword among crackpot inventors—and that's always guaranteed to raise a scientist's hackles (think of "perpetual motion" half a century earlier, or "cold fusion" half a century later).

The situation was parodied by the science fiction author Otto Binder (writing under the pen-name of Gordon A. Giles), in a story called "The Atom Smasher" in the October 1938 *Amazing Stories*. At the start of the story, a patent clerk named Milton Sander comes across a proposal for a "Basic Mass-Energy Conversion Unit":

"Atomic Power"—Sander chuckled aloud as he read those two words. . . . "Another crackpot. Perpetual Motion machines used to hold the application record, but I think lately Atomic Power engines have taken first place. When will these poor fish learn you can't get something for nothing?" [14]

In the story, of course, this particular inventor is really onto something. Correctly recognizing that the problem with Rutherford-style atom-smashers is that they only deal with one atom at time, he looks for a way to release energy from a large number of atoms at once. Binder's fictional inventor achieves this through the judicious application of sci-fi technobabble, in the form of a "gamma-ray vibro-projector"—but of course there was nothing like that in real world. Even as late as 1939, Einstein could confidently state in an interview that:

Our results so far concerning the splitting of the atom do not justify the assumption of a practical utilization of the energies released. [15]

Yet even as he spoke, the breakthrough was just around the corner. It turns out Binder's inventor was right, and you really can split trillions of atoms at the same time. All you need is a chain reaction.

Deadline

Here's a quote from a science fiction story published in John W. Campbell's magazine, *Astounding Science Fiction*, in March 1944:

Have you heard of U-235? It's an isotope of uranium. . . . U-235 has been separated in quantity easily sufficient for preliminary atomic power research, and the like. They got it out of uranium ores by new atomic isotope separation methods; they now have quantities measured in pounds. . . . But they have not brought the whole amount together, or any major portion of it. Because they are not at all sure that, once started, it would stop before all of it had been consumed—in something like one micro-microsecond of time. [16]

More will be said about this story—called "Deadline", and written by a relatively minor author named Cleve Cartmill—shortly. The thing that makes it so remarkable is that, more than a year before the first atomic bomb, it accurately describes the principle behind such bombs: the nuclear fission chain reaction.

The basic idea behind the chain reaction can be traced back several years—to 1933, when Rutherford made his much-publicized remark about nuclear power being "moonshine". One person who was set thinking by that statement was the Hungarian-born physicist Leo Szilárd, who was working in London at the time.

After reading the moonshine quote in a newspaper, it dawned on him that, in principle at least, there was a way round Rutherford's objection that atomic energy could only ever be released on microscopic scales. In Szilárd's own words:

> It suddenly occurred to me that if we could find an element which is split by neutrons, and which would emit two neutrons when it absorbed one neutron, such an element, if assembled in sufficiently large mass, could sustain a nuclear chain reaction. [17]

By March 1934, Szilárd had submitted his idea to the Patent Office. At that stage, an "idea" was all it was, since there was no known mechanism that would produce the desired chain reaction. The next step occurred in 1939, after Szilárd had moved to Columbia University in New York. Soon after he arrived there, he learned about some recent work that had been done in Germany on the use of high-energy neutrons to split the uranium nucleus—a process called "nuclear fission".

Szilárd realized what no one else had done: that nuclear fission was capable of producing a self-sustaining chain reaction. As well as releasing energy, the fission process gives rise to additional free neutrons, which can trigger the same reaction in other nuclei. If there are enough fissionable nuclei to start with—a "critical mass"—the result is exactly the sort of chain reaction Szilárd had been looking for (see Fig. 3).

Szilárd was probably the first scientist in history who, having made a great discovery, almost immediately wished he hadn't. For one thing, the large amount of energy given off by a chain reaction—all in a small fraction of a second—meant that it could be used to produce an "atom bomb" every bit as devastating as those H. G. Wells had portrayed in *The World Set Free*. Even worse, the country that led the world in fission research was Germany—which, at the time, was ruled by Adolf Hitler's aggressively militaristic Nazi regime.

Szilárd realized he needed to warn the American authorities about the situation as soon as possible. As an unknown newcomer to the country, however, there was little chance they would listen to anything he had to say about such an apparently "science-fictional" threat. So he decided to ask the most famous scientist in the United States, Albert Einstein, for assistance. At

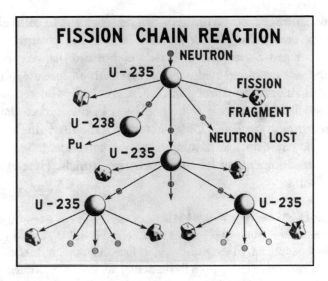

Fig. 3 Schematic diagram of a nuclear fission chain reaction in uranium (public domain image)

the time—this was the summer of 1939—Einstein was on vacation on Long Island. So that's where Szilárd headed.

One of the other scientists accompanying Szilárd on the trip was fellow Hungarian Edward Teller. That's a name that's going to pop up again and again in this book, in all sorts of different contexts. In this very first appearance he played a minor but important role, as historian Greg Herken explains:

> While their leader was Szilárd—the letter was his idea—the driving force was Teller, literally. Since Szilárd lacked an operator's licence, he relied upon Teller and the latter's temperamental 1935 Plymouth to get him to Einstein. [18]

As stated earlier, the idea of an atomic bomb was something Einstein had never taken seriously, so he was duly horrified by Szilárd's revelations. The result was a two-page letter—written by Szilárd but signed by Einstein—which was sent to President Franklin D. Roosevelt on 2 August 1939. After explaining the possibility of a powerful new bomb based on the uranium fission chain reaction, the letter concluded:

> In view of this situation you may think it desirable to have some permanent contact maintained between the administration and the group of physicists working on chain reactions in America. [19]

The immediate effect of the Einstein-Szilárd letter was to conceal all nuclear research from that point onwards under a cloak of government secrecy. Everything that had been done up to then, however, stayed in the public domain—and was devoured avidly by anyone with an interest in the subject. That included *Astounding* editor John W. Campbell, who remained just as fascinated by all things nuclear as he had been in his student days when he wrote "When the Atoms Failed". He encouraged the magazine's contributors to draw on the latest nuclear developments for story ideas, and often talked about those developments in his introductory editorials. Here is an example from July 1940:

> Assume that U-235 has been isolated in large quantities (it will be—you can bet on that). Assume that it works as physicists now believe it will. . . . . U-235 contains as much energy per pound, so the papers say, as five million pounds of coal. . . . The U-235 atom explodes, ejecting from its shattering nucleus an assortment of atomic debris including barium, tellurium, iodine, a dozen other elements, all moving outward with all the violence and velocity anyone could ask. [20]

In due course we will see what was going on beyond the public gaze, under the auspices of the top-secret Manhattan Project. Unaware of such activities, Campbell and his authors had to make educated guesses—and sometimes they were sharp enough to get pretty close to the truth. As remarkable as anything was a story Campbell printed in the May 1941 issue of *Astounding*, after World War Two had broken out in Europe, but before the United States had become involved in it. Called "Solution Unsatisfactory", the story was attributed to one "Anson MacDonald"—actually a pen-name of one of the most famous of all SF writers, Robert A. Heinlein.

Set a few years after it was written, "Solution Unsatisfactory" describes a Manhattan Project style undertaking which, in some ways, comes eerily close to the truth:

> Someone in the United States government had realized the terrific potentialities of uranium-235 quite early and, as far back as the summer of 1940, had rounded up every atomic research man in the country and had sworn them to silence. Atomic power, if ever developed, was planned to be a government monopoly, at least till the war was over. It might turn out to be the most incredibly powerful explosive ever dreamed of. . . . We were searching, there in the laboratory in Maryland, for a way to use U-235 in a controlled explosion. We had a vision of a one-ton bomb that would be a whole air raid in itself, a single explosion that would flatten out an entire industrial centre. [21]

Fig. 4 Cleve Cartmill's atom-bomb story "Deadline" appeared in the March 1944 issue of *Astounding Science Fiction* (public domain image)

Ironically, those efforts to make a "bomb that would be a whole air raid in itself"—which worked all too well in the real world—come to nothing in Heinlein's fictional version of history. In that timeline, it becomes clear by 1944 "that there existed not even a remote possibility . . . of utilizing U-235 as an explosive" [21]. In place of a bomb, the story centres around an equally deadly weapon made using radioactive dust.

The first fictional U-235 bomb had to wait for the story mentioned a few pages ago—Cleve Cartmill's "Deadline", in the March 1944 *Astounding* (see Fig. 4).

By the time "Deadline" appeared, the United States had entered the war and nuclear research was a matter of national security. To regular SF fans, however, the story wasn't a big deal—even in its technical details about the element uranium. Such matters were common currency in the SF of the time, and "Deadline" wasn't the only story to mention uranium in that particular issue. "Circle of Confusion"—written by George O. Smith under the pen-name of Wesley Long—describes how (in the story's fictional future) the terraforming of Pluto is motivated by the prospect of mining uranium there:

Pluto was found abundant in uranium, and then came Man to change the ultra-frigidity of Pluto's surface, and to endow Pluto with a breathable atmosphere. [22]

As for Cartmill's story, it was so unspectacular—or seemed so to editor Campbell—that it's relegated to the very last spot in the magazine. As SF stories go, it has a pretty poor reputation. Unlike Heinlein's "Solution Unsatisfactory", for example, it's rarely been reprinted, and Cartmill himself described it as "that stinker" [23].

In fact it would have been a good story, if Cartmill had chosen to set the action in a future extrapolation of the real world, in the same way that Heinlein did. Instead, the setting is a rather clumsily drawn alien planet, with warring alien races and no obvious connection to Earth at all. For some reason (perhaps Cartmill wanted it to read like a political satire), the alien names are all garbled versions of terrestrial ones. So, for example, the protagonist is an agent of Seilla (allies) who has infiltrated an enemy nation called Ynamre (Germany).

Why Cartmill chose to do this is a mystery. There was certainly no need, in those politically incorrect times, to avoid giving offence to people "on the other side". To see that, it's necessary to look no further than another story in the same issue—"Controller", by Eric Frank Russell. Set in the Pacific theatre of World War Two, in what would have been the present-day at the time the story appeared, it offers the reader some pretty objectionable stereotypes of Japanese soldiers.[2]

Despite its shortcomings, "Deadline" is one of the most historically important SF stories ever written. The real interest lies in its carefully explained scientific background. The protagonist's mission is to destroy the enemy's U-235 bomb before it can be used—and not just for the obvious reason of maintaining a military advantage. The Seilla scientists—i.e. the good guys—believe that once started, the nuclear chain reaction would continue unchecked until it had destroyed the whole planet (a concern that briefly cropped up in the real world, as we will see shortly).

Here is how Cartmill's protagonist describes the bomb:

> Two cast-iron hemispheres, clamped over the orange segments of cadmium alloy. And the fuze—I see it is in—a tiny can of cadmium alloy containing a speck of radium in a beryllium holder and a small explosive powerful enough to shatter the cadmium walls. Then—correct me if I'm wrong, will you?—the powdered uranium oxide runs together in the central cavity. The radium shoots neutrons into this mass—and the U-235 takes over from there. Right? [16]

---

[2] By a strange coincidence, at one point in the story a Japanese officer informs a sentry: "the password is Nagasaki" [24]. When Russell wrote that, in 1944, Nagasaki was just another Japanese city. The atom bomb wouldn't fall on it for another year.

That's not a particularly accurate description of a real atom bomb, but—with its references to neutrons and U-235—it's a cut above the usual sci-fi techno-babble. That was enough to spark an official reaction, when the story appeared on the newsstands. Both Cartmill and Campbell were promptly visited by intelligence agents, on the lookout for possible leaks from the real atom bomb project. It wasn't too hard to prove that all the story's technical details were already in the public domain—but Campbell in particular was hugely gratified to see the authorities finally taking science fiction seriously. As genre historian David Kyle explains:

> The FBI agent in Campbell's office suggested national security was being betrayed, but he was convinced finally that atom bombs and nuclear energy were common-place themes; he then proposed that such story material be eliminated. That, Campbell explained, would be suspiciously obvious and the FBI left, if not happy, at least reassured. "What I didn't show him," said Campbell, "was my circulation map with its cluster of pins for subscribers at a little place in Oak Ridge, Tennessee." Like a Geiger counter, *Astounding* and Campbell had pinpointed the hidden heart of the Manhattan Project through the scientists and technicians who read that provocative and prophetic magazine. [25]

At the time of the FBI visit, Campbell had never heard of the Manhattan Project, and he didn't know what all those *Astounding* subscribers were doing in Oak Ridge. Just over a year later, he found out—and so did the whole world.

Science Fiction Becomes Science Fact

When Robert Heinlein wrote, in his 1941 short story "Solution Unsatisfactory", that every atomic researcher in America had been rounded up and sworn to silence "as far back as the summer of 1940", he was pretty close to the truth. Soon after receiving the Einstein-Szilárd letter in the summer of 1939, the US government had set up an Advisory Committee on Uranium. Less than three years later, this had evolved into the Manhattan Project. Set up under the auspices of the US Army Corps of Engineers, it was originally based at Tower 270 in downtown Manhattan—leading to its formal designation as the "Manhattan Engineer District".

The Manhattan Project's task was twofold: to design a viable atomic bomb, and to acquire enough fissionable material to construct it. Both were

Fig. 5 The birth of the Manhattan Project, from an *Adventures in Science* comic book produced by the General Electric Company in 1948 (public domain image)

formidable challenges, and the result was the biggest team effort the scientific world had ever seen, spread over multiple sites across the United States (see Fig. 5).

The Manhattan Project's lead scientist was a 38-year-old California professor named Robert Oppenheimer. At the time, he was almost as famous for his left-wing political views as for his scientific abilities. Being a bit "red", however, wasn't the show-stopping problem it would later become—since the enemies of the day were Nazis, not communists.

One of the most striking things about the Manhattan Project—particularly in view of its relation to "national" security—is just how *inter*national it was. For example, two of its leading figures who have already been mentioned—Leo Szilard and Edward Teller—were both Hungarian. Also present were Nobel-prizewinning scientists from Italy (Enrico Fermi), Denmark (Niels Bohr),

Britain (James Chadwick) and America (Ernest Lawrence). The latter was the most recent prizewinner—in 1939, for his invention of a new kind of particle accelerator called the cyclotron.

The list of famous Manhattan Project scientists goes on and on. To mention just those whose names are going to crop up later in this book—there's John von Neumann, another Hungarian; Stanisław Ulam, from Poland; Hans Bethe, from Germany, and another American, Luis Alvarez (the last two also became Nobel prizewinners after the war).

Despite the vast intellectual power the government threw at the Manhattan Project, one of its most challenging problems was a practical rather than a scientific one—the extreme scarcity of fissionable material. Naturally occurring uranium consists almost entirely of the stable isotope U-238, with fissionable U-235 accounting for less than one per cent. To get a self-sustaining chain reaction requires a "critical mass" of U-235, which even in pure form amounts to more than 50 kg of it.

This meant that U-235 had to be separated out of natural uranium—something that could be achieved using a version of Lawrence's cyclotron, but only in a very slow and laborious process. This was what was being done at Oak Ridge in Tennessee—the place where John W. Campbell noticed all those *Astounding* subscribers suddenly moved to.

The initial focus was on U-235 because it's a naturally occurring substance, albeit in small quantities. But in 1940 another fissile isotope was discovered—plutonium-239—which in some ways was more suitable for weapons applications. Although it doesn't occur naturally, plutonium can be created in a nuclear reactor from the relatively common isotope U-238—making it ultimately quicker to produce than U-235. So another Manhattan Project site was set up—at Hanford in Washington state—for this purpose.

Hanford and Oak Ridge can be considered the dull and dirty side of the Manhattan Project, slowly churning out the necessary materials for a bomb. The cutting edge scientific research—the design of the bomb itself—took place at yet another site, at Los Alamos in New Mexico.

Eventually two bomb designs were settled on. The simpler option, called a "gun-type" device, involves a slightly sub-critical chunk of U-235 being fired—gun-style, as the name suggests—into a second sub-critical chunk. Since the two chunks together are now greater than the critical mass, by quite a margin, a successful chain reaction is virtually certain. If anything, it's too certain. A bomb of this type could easily go off if it was dropped or mishandled, making it almost as dangerous to its operators as the enemy.

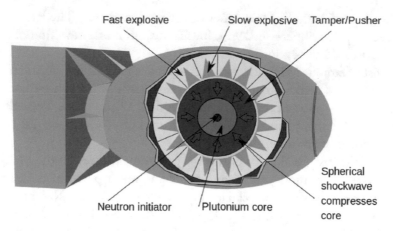

Fast explosive   Slow explosive   Tamper/Pusher

Neutron initiator   Plutonium core   Spherical shockwave compresses core

Fig. 6 Simplified diagram of a "Fat Man" implosion-type atomic bomb (Wikimedia user Fastfission, CC-BY-SA-2.5)

The other design that came out of the Manhattan Project is much more sophisticated. Called an "implosion-type" device, it uses plutonium as the fissile material—but only about 40% of the critical mass. You could drop it or play with it to your heart's content and it wouldn't explode until you wanted it to. To achieve that, conventional explosives are used to compress (or "implode") the plutonium to a state where it's dense enough to go critical. That's much safer than a gun-type device—and also more efficient, because it uses less fissile material to the same effect.

The implosion-type bomb became known as "Fat Man" (see Fig. 6)—a name suggested by one of Oppenheimer's assistants, after the character played by Sidney Greenstreet in the 1941 film noir *The Maltese Falcon* [26].

The world's first atomic explosion came with the test of a Fat Man bomb on 16 July 1945 (the gun-type design, called "Little Boy", was so simple—and so wasteful of U-235—that it wasn't given a test). Code-named Trinity, the Fat Man test took place at the White Sands test range in New Mexico. It generated an explosive energy equivalent to about 20,000 tons of TNT—or "20 kilotons", in Manhattan Project jargon. The result was an 80-m crater in the desert sand, and an explosion heard more than 150 km away.

By this time, Nazi Germany had been defeated and the focus of America's war effort had shifted to Imperial Japan. A war of attrition was already being fought against Japanese cities, using giant B-29 bombers flying from bases in the Mariana Islands, about 2000 km south of Japan in the Pacific Ocean. Four months before the Trinity test, Operation Meetinghouse had already seen the single most devastating air raid in history. On the night of 9–10 March 1945, around 300 B-29s dropped thousands of bombs on Tokyo, killing almost a hundred thousand people.

A month after Operation Meetinghouse, President F. D. Roosevelt died, to be succeeded by his Vice President, Harry S. Truman—who only then became aware of the existence of the Manhattan Project. He realized it was the ace up his sleeve that could end the war against Japan. Soon after the Trinity test, Truman attended a summit meeting in Potsdam, near Berlin. On 26 July 1945, the conference issued an ultimatum to Japan—the Potsdam Declaration—which ended with the words:

> We call upon the government of Japan to proclaim now the unconditional surrender of all Japanese armed forces, and to provide proper and adequate assurances of their good faith in such action. The alternative for Japan is prompt and utter destruction. [27]

Since no explanation was given as to how this "prompt and utter destruction" would be achieved, the Japanese government didn't take the threat seriously. In the background, however, things were happening. A special unit of the US Army Air Force, the 509th Composite Group, had been deployed to Tinian Island in the Marianas, equipped with specially modified B-29 bombers. Codenamed Silverplate, these aircraft really were capable of delivering "prompt and utter destruction".

On the day of the Potsdam Declaration a number of crates were delivered to Tinian airfield, containing the bomb casing, internal mechanism and fissile core of the world's first combat-ready atom bomb. This was a "Little Boy" gun-type device—and it was joined a few days later by a second bomb of the more sophisticated "Fat Man" type. With Little Boy and Fat Man on the scene, all that remained was to pick out some targets.

Three possible targets were identified: the cities of Hiroshima, Kokura and Nagasaki. The last of these was best known in the west as the setting of Puccini's 1904 opera *Madame Butterfly*—which, incidentally, is the most likely reason Eric Frank Russell used it as the Japanese password in his story "Controller", which appeared alongside Cleve Cartmill's "Deadline" in the March 1944 issue of *Astounding Science Fiction*.

Hiroshima was selected as the target for the first mission, on 6 August 1945, and three Silverplate aircraft were duly dispatched. One of them delivered the bomb itself—Little Boy on this occasion—while the other two made various scientific measurements. Three days later it was Fat Man's turn. This time the primary target was Kokura, but when the strike package arrived there the

Fig. 7 Manhattan Project scientist Luis Alvarez in front of a Silverplate B-29 at Tinian airfield in 1945 (public domain image)

weather was too cloudy for the crew to aim the bomb, so the target was switched at the last minute to Nagasaki.

Although only two bombs were dropped—compared to thousands in the case of the Operation Meetinghouse air raid on Tokyo—each of them proved as destructive as the entire Tokyo raid. Both target cities, Hiroshima and Nagasaki, were almost completely destroyed, and the death toll continued to rise over time due to the lingering effects of radiation. Casualty estimates vary widely, but may have been as high as 144,000 (48% of the total population) in Hiroshima and 59,000 (30%) in Nagasaki [28].

On board one of the scientific support aircraft for the Hiroshima mission was Manhattan Project scientist Luis Alvarez (see Fig. 7). Aware of all the painstaking effort the Oak Ridge team had put into uranium separation, his first impression on witnessing the aftermath of the bomb was that it had all gone to waste:

> I looked out and all I could see was a black, roiling cloud over what looked like a forest. My first thought to myself was that Ernest Lawrence would be furious when he learned that they had wasted all his uranium on a forest. I didn't see any sign of a city. [29]

Like so many people who worked on the Manhattan Project, Alvarez was a hugely important scientist in his own right. His work on elementary particle physics won him the Nobel Prize in 1968, and in the 1980s he helped originate the theory that a giant meteorite wiped out the dinosaurs. Alvarez was also a friend and one-time colleague of SF writer Arthur C. Clarke, who remarked that "Luis seems to have been there at most of the high points of modern physics—and responsible for many of them" [30]. Whether the bombing of Hiroshima counts as a "high point" of physics is debatable—but it was certainly an important one.

When President Truman announced the bombing of Hiroshima, later that same day, he too made the point that it represented a victory—of a kind—for science:

We have spent two billion dollars on the greatest scientific gamble in history—and won. But the greatest marvel is not the size of the enterprise, its secrecy, nor its cost, but the achievement of scientific brains in putting together infinitely complex pieces of knowledge held by many men in different fields of science into a workable plan. . . . What has been done is the greatest achievement of organized science in history. [31]

The advent of the atomic bomb had enormous repercussions in every sphere—military, scientific, political and cultural. Of particular relevance as far as this book is concerned—it transformed the public's attitude toward science fiction almost overnight. All of a sudden, SF looked a lot less fanciful and irrelevant than it had a few years earlier. This was something John W. Campbell emphasized in one of his first post-war editorials, in the November 1945 issue of *Astounding*:

The atomic bomb fell, and the war was, of course, ended. During the weeks immediately following that first atomic bomb, the science-fictioneers were suddenly recognized by their neighbours as not quite such wild-eyed dreamers as they had been thought, and in many soul-satisfying cases became the neighbourhood experts. [32]

Many of the less sensitive details of the Manhattan Project—both technical and administrative—were made public immediately after the end of the war in an official report by Henry D. Smyth. Campbell chose to reprint a section of this report in the December 1945 *Astounding*, prefacing it with the following words:

This excerpt from the Smyth Report discusses the last stage of the problem of producing the atomic bomb; the mechanism whereby the uranium isotope U-235, or the synthetic element plutonium, Pu-239, could be detonated. May we call to your attention that the essential principles employed in the bomb were described as the arming mechanism of the atomic bomb in the story "Deadline" in the March, 1944 *Astounding*. [33]

Actually, it's stretching things to suggest that Cartmill's story provides a clear description of a real atom bomb. Even if it wasn't a bull's-eye, however, it was certainly a near-miss. The same could be said of the story's most grandiosely over-the-top element—the idea that a runaway chain reaction might end up destroying a whole planet.

As archetypally "sci-fi" as this sounds, something similar really was considered in the course of the Manhattan Project—by a character we're going to hear much more about in this book, Edward Teller. As his fellow Manhattan Project scientist Hans Bethe explained in 1991, recalling events that took place in 1942:

One day Teller came to the office and said, "Well, what would happen to the air if an atomic bomb were exploded in the air?... There's nitrogen in the air, and you can have a nuclear reaction in which two nitrogen nuclei collide and become oxygen plus carbon, and in this process you set free a lot of energy. Couldn't that happen?" [34]

Teller's fear was that once the carbon-nitrogen reaction got started, it would carry on until it had burned up the whole atmosphere. A few calculations quickly showed this to be impossible—but the idea wasn't forgotten. One of the other scientists, Enrico Fermi, brought it up in a light-hearted way on the occasion of the Trinity test, near Los Alamos in 1945. Quoting Bethe again:

Fermi, of course, didn't believe that this was possible, but just to relieve the tension at the Los Alamos test, he said, "Now, let's make a bet whether the atmosphere will be set on fire by this test." [34]

The Start of the Arms Race

The immediate public reaction to the atom bomb, before the horrific effects of radiation became widely known, wasn't as negative as it might have been. After all, it had brought the long-drawn-out war against Japan to a sudden end, and reinforced America's self-image as the most technologically advanced nation

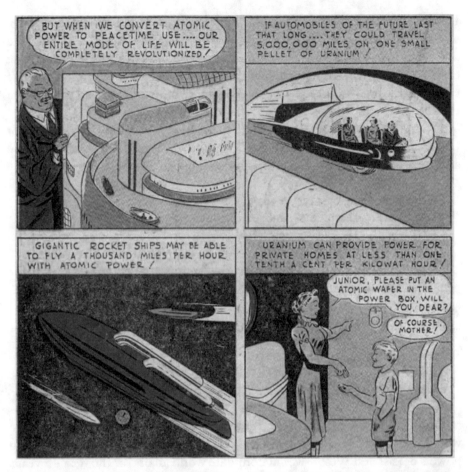

Fig. 8 A few panels from an optimistic piece about "The Atomic Age", from the second issue of *Marvels of Science*, published by Charlton Comics in April 1946 (public domain image)

on Earth. When the first post-war nuclear test took place in July 1946, it wasn't something people viewed as bleak or threatening. Quite the opposite, in fact: the location of the test, a small atoll in the Pacific called Bikini, soon gave its name to a trendy new item of summer beachwear.

There was another reason for the atomic optimism of the immediate post-war period: the realization that nuclear power could also be used for peaceful purposes. In the popular imagination, the word "atomic" became an excitingly futuristic adjective, heralding all sorts of possibilities for the clean, efficient "Atomic Age" to come (see Fig. 8).

Unfortunately there was a flaw in this optimistic vision of the future—and it had nothing to do with science or technology, but with politics. The Soviet

Union, which had been a staunch ally of the West during World War Two, was set to become its number one enemy. Basically it came down to a difference in ideology: Soviet communism versus American-style free market capitalism. In scientific matters, however, the Soviets were every bit America's equals—and that meant they were capable of building nuclear weapons of their own.

One of the first people to see where the world was heading was the author George Orwell. As early as October 1945, he wrote an article entitled "You and the Atomic Bomb" for the British newspaper *Tribune*:

> From various symptoms one can infer that the Russians do not yet possess the secret of making the atomic bomb; on the other hand, the consensus of opinion seems to be that they will possess it within a few years. [35]

Of course, many people could have said the same thing—but Orwell's genius was to see what that meant for the future:

> Had the atomic bomb turned out to be something as cheap and easily manufactured as a bicycle or an alarm clock, it might well have plunged us back into barbarism. . . .. If, as seems to be the case, it is a rare and costly object as difficult to produce as a battleship, it is likelier to put an end to large-scale wars at the cost of prolonging indefinitely a "peace that is no peace". [35]

Orwell went on to describe this situation as "a permanent state of cold war"—and the term stuck. The Cold War—between the two great nuclear powers of East and West—would dominate world affairs for the next 45 years.

The idea of a Cold War lies at the heart of George Orwell's most famous work—the 1949 novel, *Nineteen Eighty-Four*. As journalist David Aaronovitch explains:

> It's often missed that *Nineteen Eighty-Four* is set a few decades after an atomic war. The managers administering the book's three super-states . . . have tacitly agreed not to try to destroy each other but to continue forever in a kind of cold war. [36]

Here is a passage from the novel itself which makes this clear:

> Atomic bombs first appeared as early as the 1940s, and were first used on a large scale about ten years later. At that time some hundreds of bombs were dropped on industrial centres, chiefly in European Russia, Western Europe and North America. The effect was to convince the ruling groups of all countries that a few

more atomic bombs would mean the end of organized society, and hence of their own power. Thereafter, although no formal agreement was ever made or hinted at, no more bombs were dropped. All three powers merely continue to produce atomic bombs and store them up against the decisive opportunity which they all believe will come sooner or later. [37]

The real Cold War was even "colder" than Orwell imagined, without even a limited nuclear exchange in the 1950s. There's a reason for that, and it's something Orwell, writing in the 1940s, failed to anticipate. He was thinking in terms of the kiloton-class atom bombs that were dropped on Hiroshima and Nagasaki, each of which was equivalent to a very heavy air raid by conventional bombers.

But what if there was an even bigger bomb—a thousand times bigger? Its power would be measured not in kilotons but megatons, and rather than being an air raid in itself, a single explosion would be a whole war in itself. With more experience in thinking about technological futures than Orwell, someone who did foresee such a super-bomb was John W. Campbell, editor of *Astounding Science Fiction*. As he pointed out in an editorial in January 1946:

The uranium reaction is reasonably potent, but another one, discovered in 1930 by Lord Rutherford, is nearly twice as powerful, pound for pound, and uses cheap lithium and ordinary hydrogen. It won't start until a temperature of several million degrees is reached, but the Hiroshima U-235 bomb would make an excellent primer to start the more violent explosion. [38]

As with Campbell's earlier writings on the subject of uranium fission, he hit the nail on the head. What he's talking about here is a different type of nuclear reaction—not fission, but fusion. It's the reaction that goes on inside the Sun, and it can only take place at extremely high temperatures—for which reason it's often referred to as a "thermonuclear" reaction.

A few years earlier, in the winter of 1941, this same principle had cropped up in a conversation between two of the world's top nuclear physicists, Enrico Fermi and Edward Teller. To quote historian Gregg Herken:

Over lunch, Fermi had made the simple observation that an atomic bomb might release enough energy to start the thermonuclear reaction that fuelled the Sun and other stars. Powerful as a fission bomb might be, it would then be only an initiator for a very much larger bomb of a different sort—fusing hydrogen into helium. A so-called hydrogen bomb could theoretically be of virtually unlimited power. [39]

Teller, for one, was hugely excited by the idea of a hydrogen bomb—or H-bomb, as it became known. While many of his fellow scientists wanted nothing to do with such a devastating weapon, Teller relentlessly pushed for America to build one before the Soviets got there first. As he put it in 1945: "If the development is possible, it is out of our powers to prevent it" [40].

In a sense, Teller was right. Soviet scientists could envisage an H-bomb as well as Americans could, and the race was on to build one whether they liked it or not. While the American side prevaricated—over everything from technical feasibility and military necessity to cost-effectiveness and ethics—the Soviets quietly crept ahead. In the words of military historian David Baker:

> It is now clear that the determination of the Russians to move quickly from a fissile bomb to a fusion device had a degree of seamless inevitability lacking in the decision of the United States to acquire a thermonuclear weapon. [41]

The first bomb to incorporate fusion effects wasn't a fully fledged thermonuclear weapon, but something called a "boosted fission device"—essentially a Trinity-style Fat Man bomb, in which a small amount of additional energy was provided by fusion. The first American test of such a device came in May 1951, six years after the Trinity test. Just four months later—and only two years after the first Soviet atom bomb—the Russians tested their own boosted fission weapon. The Americans might be leading the race, but it was the Russians who were running faster.

The power of a boosted fission weapon was still only a matter of kilotons, not megatons. Teller was still grappling with the problem of how to build a true, megaton-class H-bomb. The breakthrough came when he saw a design by fellow physicist Stanisław Ulam, using a fission bomb to initiate a full-scale fusion reaction. As Greg Herken explains:

> Ulam's sketch inspired Teller to think of another, better way to compress the fusion fuel without heating it—by using radiation from the exploding fission trigger, travelling at the speed of light. [42]

The resulting Teller-Ulam design was tested in November 1952, under the codename Ivy Mike. This was by far the most awesome man-made explosion to date—at more than 10 megatons, it was around 40 times more powerful than the boosted fission test the previous year.

There was a snag, though. As successful as the Ivy Mike test was, it wasn't a "bomb" by any stretch of the imagination. It was a huge, cumbersome scientific test rig, using cryogenically cooled liquid hydrogen as the fusion fuel. It wasn't the sort of thing that could ever be used as a weapon—which required a small, compact design employing "dry" fuel rather than a liquid.

The world first true H-bomb exploded on 22 November 1955. It had a yield of three megatons and it wasn't just a bulky test rig—it was a small, neat package dropped from an aircraft. The bomb detonated over the Semipalatinsk test site in the Soviet Union, and the aircraft dropping it had a red star painted on its tail. For the first time in history, the United States was losing the nuclear arms race. As aerospace engineer—and occasional science fiction author—G. Harry Stine wrote:

> It comes as a real shock to most people to realize that the Soviets beat the United States to the thermonuclear warhead. No one can properly consider the American thermonuclear test, Shot Mike, as a warhead or even a weapon; at best it was a complex, bulky, building-sized, 65-ton proof-of-principle test gadget. It showed the United States that a thermonuclear reaction could be triggered but that it would have to be made smaller, simpler and lighter to be considered as a weapon. . ... On the other hand, the Soviets went directly to the dry thermonuclear weapon. Unlike the situation with the atomic bomb where the USSR reacted to what the United States had done, it was the United States that reacted to the Soviet achievement. [43]

Nuclear Paranoia

The advent of the H-bomb, and the Cold War arms race that followed, shaped the world-view of a whole generation. Suddenly everyone in America was paranoid—though not necessarily about the same thing. Left-wingers were terrified by the thought that a nuclear war might break out against the Soviet Union at any moment—an eventuality they wanted to avoid at any cost ("better Red than dead"). At the same time, right-wingers were equally terrified by the encroaching threat of communism itself—again, something they wanted to prevent at any cost ("better dead than Red").

The latter view was spearheaded by Senator Joseph McCarthy, who became convinced that communists and Soviet sympathizers had infiltrated all areas of US society. The result was a notorious "witch hunt", which saw anyone who dared to express left-wing opinions branded as disloyal to the United States. The House Un-American Activities Committee (HUAC) had existed since before World War Two, but under McCarthy's influence it really took off in the 1950s.

HUAC was responsible for destroying countless careers—including those of many scientists. One high-profile victim was the former head of the

Manhattan Project, Robert Oppenheimer, who lost his security clearance in 1954. Perhaps that was inevitable, given Oppenheimer's well-known liberal-ism—but it's still a shocking fate for the man who gave America its war-winning weapon.

Also on the receiving end of HUAC's boot was the Chinese-born rocket scientist Hsue-Shen Tsien (we'll meet him again in the next chapter, "Journey into Space"). Arthur C. Clarke gave Tsien the honour of naming a spacecraft after him in his novel *2010: Odyssey Two* (1982). As Clarke explains in the introduction to that book:

> He was the first Goddard Professor at the California Institute of Technology, and contributed greatly to American rocket research through the 1940s. Later, in one of the most disgraceful episodes of the McCarthy period, he was arrested on trumped-up security charges when he wished to return to his native country. For the last two decades, he has been one of the leaders of the Chinese rocket programme. [44]

HUAC was a mind-blowing example of a country wilfully shooting itself in the foot. At the very moment America was locked in a life-or-death techno-logical race with the Soviets, the witch-hunts surgically removed some of the key people who could have helped to keep it ahead in that race.

As it was, America steadily fell behind—and the blame for that didn't lie solely with the right-wingers. At the other political extreme, the existence of a vocal anti-nuclear movement—a problem that Russia never suffered from—meant that America was being pulled in two directions at once.

The same left-right polarization was evident in the science fiction of the time. As social historian S. D. Tucker points out, with regard to "the idea that the West should unilaterally surrender its nuclear weapons":

> Hollywood was at the forefront of spreading such a message. Most famously, there was the 1951 movie *The Day the Earth Stood Still*, in which human-looking Space Brother Klaatu and his invincible robot Gort threaten the planet with destruction if it doesn't renounce its atom-splitting ways immediately. [45]

But for every left-wing anti-nuclear sci-fi film, there was a right-wing anti-communist one. To quote Tucker again:

> The other side of the coin, meanwhile, could be seen in paranoid classics like 1953's *Invaders from Mars* and 1955's *Invasion of the Body Snatchers*, in which the aliens become metaphors for potential Commie infiltrators during the age of McCarthyism. [45]

While the American public argued with itself over the rights and wrongs of the arms race, the Russians—unfettered by such worries—crept further and further ahead. Both sides were conducting increasingly sophisticated nuclear tests, but the balance looked to be in favour of the Soviets. By 1958, many people felt it would be in America's interest to impose a global moratorium on such testing. Not everyone agreed with this view, though. Among the voices raised against it were those of SF author Robert A. Heinlein and his wife Virginia. As Heinlein biographer James Gifford explains:

In 1958, President Eisenhower was considering a unilateral cessation of nuclear weapon testing, based on a Soviet promise to make it joint. The Heinleins were adamantly opposed, given the Soviet Union's poor record of promise-keeping .... Shortly thereafter, the Soviet Union ignored its promise and resumed testing with some of the largest and dirtiest weapons ever detonated. Heinlein was infuriated. He stopped work on the novel that would become *Stranger in a Strange Land* and wrote *Starship Troopers* in a white-hot fury. [46]

Starship Troopers turned out to be one of Heinlein's most successful works, winning the prestigious Hugo Award for best novel of the year. Before it was published in book form, an abridged version was serialized in *The Magazine of Fantasy & Science Fiction* as "Starship Soldier" (see Fig. 9).

Starship Troopers is set against the backdrop of a future space war, with humans pitted against alien "Bugs" who are thinly veiled stand-ins for Cold War communists:

They are arthropods who happen to look like a madman's conception of a giant, intelligent spider, but their organization, psychological and economic, is more like that of ants or termites; they are communal entities, the ultimate dictatorship of the hive. [47]

Unlike virtually every other SF novel of the 1950s, *Starship Troopers* has far more talk than action. To make things even more controversial, much of that talk is politically motivated—a long rant against the spread of liberalism in American society. Heinlein's fellow author Anthony Boucher—who had been the editor of *Fantasy & Science Fiction* himself at one time—wrote a scathing review of the book for the *Herald Tribune*:

This is not a novel at all, but an irate sermon with a few fictional trappings. Mr Heinlein, an angry middle-aged man, wishes to denounce the decadence of mid-20th century America and to advocate a more Spartan civilization. ... The author is so intent upon his arguments that he has forgotten to insert a story. [48]

Fig. 9 The cover of the November 1959 issue of *The Magazine of Fantasy & Science Fiction*, featuring the first instalment of Robert A. Heinlein's "Starship Soldier"—later to become *Starship Troopers* (public domain image)

More charitably, one could say that *Starship Troopers* was simply a few years ahead of its time. By the mid-1960s, technology-focused SF was falling out of fashion in favour of a stronger emphasis on social commentary. Ironically, however, most of that later "sociological" SF lay on the political left, while Heinlein's novel is emphatically right-wing. He argues that a strong military is essential—not, as pacifists imagine, simply for the sake of killing people, but as the only practical way for a country to maintain a robust position on the international stage:

> War is controlled violence, for a purpose. The purpose of war is to support your government's decisions by force. The purpose is never to kill the enemy just to be killing him, but to make him do what you want him to do. [49]

Starship Troopers is one of the few SF novels of the mid-20th century that remains popular to this day—and not just with the general public. The US Marine Corps includes it on their official reading list for entry-level recruits and officer candidates [50]. Bearing in mind that Heinlein was motivated to write the novel by a now-forgotten political situation—effectively as a piece of anti-communist propaganda—it's even more remarkable that it's outlived the times in which it was written.

On the other hand, it's no surprise at all that the publication of *Starship Troopers* had no impact whatsoever on the situation that motivated it—the relentless tide of Soviet nuclear testing. That reached its peak on 30 October 1961, with the detonation of the 50-megaton Tsar Bomba. Even to this day, it remains the most powerful bomb ever built—and it was a tactically viable weapon, dropped from an aircraft rather than being exploded on a stationary test rig. It was a graphic demonstration that the Soviet Union, and not the United States, was now the world's number one nuclear superpower. As David Baker put it:

> The effect of the Tsar Bomba worked its trick. The bomb had been dropped over the Novaya Zemlya test site and it had been spectacular. Released at a height of 10,300 m and detonated at an altitude of 4,000 m six minutes later … it was observed 1,000 km away by the State Commission, a distance necessitated because the blast alone would cause third degree burns at a distance of 100 km. All the buildings, including those made of brick, were totally destroyed in a village 55 km away and a spectator wearing goggles felt the heat from the thermal energy 270 km distant. … By lensing through the atmosphere, windows were broken in Finland and Norway and the shock wave that hit the ground went round the world three times. [51]

The Tsar Bomba was the first weapon of truly sci-fi proportions. Yes, the Hiroshima bomb caused terrible suffering and loss of life, but the total destruction wasn't significantly greater than the conventional Meetinghouse air raid a few months earlier—and very few people have heard of that. They've heard of Hiroshima because of the horrifying fact that it was a single weapon, dropped by a single bomber. So was the Tsar Bomba—and it was 2500 times more powerful. That's so utterly, indiscriminately destructive that it's no longer a militarily useful weapon. It's a mad scientist's weapon—or a mad politician's.

Pandora's Box

The existence of weapons like the Tsar Bomba was a nightmare from which the world couldn't wake up. Once something has been invented, it can't be uninvented—as G. Harry Stine points out:

> Nuclear warhead technology is a Pandora's box; it has been opened and is loose in the world. No way exists to put it back in the box and hold the lid closed. No restrictions on technology transfer can keep the know-how within national borders. Once an engineer knows that another engineer has done something, said engineer can proceed with the assurance that he, too, can do it because the laws of the universe are the same everywhere. [52]

Stine's analogy with the legend of Pandora's Box is an apt one. Another way to put it would be to say "you can't put the genie back in the bottle"—and the two metaphors are closely related, as Brian Clegg points out:

> In ancient Greek mythology, Pandora was the equivalent of Eve, the first woman, created directly by the gods. She was given by her creators a jar that was never to be opened, containing all the ills of the world. ... The jar was, in Greek, called a *pithos*, from which we probably get the word "pitcher". ... But when the poem was translated to Latin by the medieval Dutch scholar Erasmus, he mistakenly turned *pithos* into *pyxis*, which is Latin for "box". So the original Pandora myth was closer to the Arabic stories where evil djinns are locked away in jars. [53]

There was something different about the nuclear Pandora's box, however. The sheer destructiveness of H-bombs led to a worldwide reluctance to use them—and quite possibly, as Orwell had prophesied in 1945, helped to prevent an all-out war breaking out between East and West.

That didn't mean people were happy with the situation, and both sides quietly sought a way out of the impasse. The SF writer George O. Smith described their dilemma at the start of his story "The Undamned", in the January 1947 issue of *Astounding* magazine:

> Plutonium was an equalizer. Nations learned the art of being polite ... to lash out with plutonium wildly would be inviting national disaster, and to behave in an antisocial manner would get any nation the combined hatred of the rest of the world—equally a national disaster. This was surface politeness. Beneath, the work went on to find an adequate defence, for now that all nations were equal, the first capable of defending itself was to be winner. [54]

In the story, a solution to the problem is found in the form of a defensive force-field that renders nuclear weapons impotent. The same idea cropped up in a slightly later, and much better known, *Astounding* story: Isaac Asimov's "Breeds There a Man?", from the June 1951 issue. In one scene, Asimov has a physicist explain the concept to a medical doctor:

"So far, military advances have been made fairly equally in both offensive and defensive weapons. Once before there seemed to be a definite and permanent tipping of all warfare in the direction of the offence, and that was with the invention of gunpowder. But the defence caught up. The mediaeval man-in-armour-on-horse became the modern man-in-tank-on-treads, and the stone castle became the concrete pillbox. The same thing, you see, except that every-thing has been boosted several orders of magnitude."

"Very good. You make it clear. But with the atomic bomb comes more orders of magnitude, no? You must go past concrete and steel for protection."

"Right. Only we can't just make thicker and thicker walls. We've run out of materials that are strong enough. So we must abandon materials altogether. If the atom attacks, we must let the atom defend. We will use energy itself; a force field."

"And what," asked Blaustein, gently, "is a force field?"

"I wish I could tell you. Right now, it's an equation on paper. Energy can be so channelled as to create a wall of matterless inertia, theoretically. In practice, we don't know how to do it."

"It would be a wall you could not go through, is that it? Even for atoms?"

"Even for atom bombs. The only limit on its strength would be the amount of energy we could pour into it. It could even theoretically be made to be imper-meable to radiation. The gamma rays would bounce off it. What we're dreaming of is a screen that would be in permanent place about cities; at minimum strength, using practically no energy. It could then be triggered to maximum intensity in a fraction of a millisecond at the impingement of short-wave radiation; say the amount radiating from the mass of plutonium large enough to be an atomic war head. All this is theoretically possible." [55]

That, of course, would be the perfect solution to everyone's nuclear worries. Sadly, however, a protective force-field isn't even "theoretically possible" in the real world—only in the pages of science fiction.

In the absence of any viable defence, much of the Cold War was dominated by a bizarre concept known as Mutual Assured Destruction, or MAD for short. To quote military historian Pat Ware:

In the interests of maintaining the uneasy stalemate, it was essential that neither side gained a technological advantage. However, even though both sides might have equal firepower, should one side decide to launch a comprehensive attack on the other there remains a window of opportunity for the other side to react before being annihilated. This ability to react, even after having come under attack—a second strike—forms an intrinsic part of a strategy that came to be described as "Mutual Assured Destruction"—a term which yields the ironically appropriate acronym MAD.

Both the strategy, and the acronym, came from the American scientist John von Neumann who was chairman of the International Ballistic Missile Committee. MAD ensured that any nation that had come under attack would be able to respond with a nuclear retaliation of at least equal force. [56]

Like Edward Teller and Leo Szilárd, John von Neumann was a Hungarian-born scientist who worked on the Manhattan Project. MAD wasn't the only idea he came up whose frivolous-sounding name belied its intensely serious nature. Another of his inventions, "game theory", was anything but a game—as we shall see in the third chapter, "Electronic Brains". As for MAD itself—the acronym may have been von Neumann's, but the underlying concept wasn't entirely original to him. At any rate, Arthur C. Clarke—in his "science-fictional autobiography" *Astounding Days*—laid claim to it:

In 1946, my *Royal Air Force Quarterly* essay "The Rocket and the Future of Warfare" explored all the possibilities opened up by the advent of the V-2 and the atomic bomb. It ended with this paragraph:

"One returns to the conclusion that the only defence against the weapons of the future is to prevent them ever being used. In other words, the problem is political and not military at all. A country's armed forces can no longer defend it: the most they can promise is the destruction of the attacker."

To have invented MAD is not one of my prouder achievements. [57]

The MAD stalemate led to the surreal situation in which a genuine global conflict—the Cold War—was fought not on battlefields but in scientific research establishments, where each side desperately tried to out-think the other. The result was a long-drawn-out but uneasy peace—a situation that was acceptable to the inhabitants of Earth, but perhaps less so to certain extraterrestrials. It would, for example, have been very frustrating for any aliens who happened to be waiting, vulture-like, for a nuclear war to break out so they could come in and take over. That's the basic scenario in Isaac Asimov's 1957 short story "The Gentle Vultures". As one alien complains to another:

Fig. 10 A panel from the first issue of a comic-book called *Atom Age Combat*, from February 1958—"For the sake of global survival, it is the common duty of all of us to work incessantly to prevent the suicidal thermonuclear holocaust of total war in the Atom Age!" (public domain image)

> This planet has something called a Cold War. Whatever it is, it drives them furiously onward in research and yet it does not involve complete nuclear destruction. [58]

As annoying as that may have been for Asimov's aliens, it could only be a good thing for the people of Earth. The Cold War may have been the only period of history in which virtually everyone, in their own way, became a pacifist. To those on the left, the answer was disarmament; to those on the right, it was eternal vigilance. To use the Cold War's favourite cliché, nuclear conflict had become "unthinkable"—and the popular media of the time never let people forget that (see Fig. 10).

References

1. I. Asimov, *Biographical Encyclopedia of Science and Technology* (Pan Books, London, 1975), pp. 11, 12
2. I. Asimov, *Biographical Encyclopedia of Science and Technology* (Pan Books, London, 1975), p. 232
3. B. Clegg, *Armageddon Science* (St. Martin's Press, New York, 2010), p. 165

4. P. Bizony, *Atom* (Icon Books, London, 2007), p. 11
5. P. Nicholls, *The Science in Science Fiction* (Book Club Associates, London, 1983), p. 32
6. H.G. Wells, *The World Set Free* (Project Gutenberg, 2006), http://www.guten berg.org/ebooks/1059
7. A. Christie, *The Big Four* (Fontana, London, 1965), p. 147
8. A. Christie, *The Big Four* (Fontana, London, 1965), p. 157
9. J.W. Campbell, When the Atoms Failed, in *Amazing Stories* (January 1930), pp. 910–925
10. A. Einstein, *Ideas and Opinions* (Souvenir Press, London, 2005), pp. 339, 340
11. W. Isaacson, *Einstein: His Life and Universe* (Pocket Books, London, 2008), p. 272
12. P. Bizony, *Atom* (Icon Books, London, 2007), p. 88
13. B. Clegg, *Armageddon Science* (St. Martin's Press, New York, 2010), p. 41
14. O. Binder [as Gordon A. Giles], The Atom Smasher, in *Amazing Stories* (October 1938), pp. 30–41
15. W. Isaacson, *Einstein: His Life and Universe* (Pocket Books, London, 2008), pp. 469, 470
16. C. Cartmill, Deadline, in *Astounding Science Fiction* (March 1944), pp. 154–178
17. B. Clegg, *Armageddon Science* (St. Martin's Press, New York, 2010), p. 42
18. G. Herken, *Brotherhood of the Bomb* (Owl Books, New York, 2003), p. 24
19. W. Isaacson, *Einstein: His Life and Universe* (Pocket Books, London, 2008), p. 474
20. J.W. Campbell, *Astounding Science Fiction* (July 1940), pp. 5, 6
21. R.A. Heinlein, Solution Unsatisfactory, in *The Worlds of Robert A. Heinlein* (New English Library, London, 1970), pp. 92–127
22. G.O. Smith [as Wesley Long], Circle of Confusion, in *Astounding Science Fiction* (March 1944), pp. 45–65
23. A. Rogers, *A Requiem for Astounding* (Advent, Chicago, 1964), pp. 132, 133
24. E.F. Russell, Controller, in *Astounding Science Fiction* (March 1944), pp. 66–85
25. D. Kyle, *A Pictorial History of Science Fiction* (Hamlyn, London, 1976), p. 113
26. G. Herken, *Brotherhood of the Bomb* (Owl Books, New York, 2003), p. 84
27. The Potsdam Declaration (26 July 1945), *The Atomic Archive*, http://www.atomicarchive.com/Docs/Hiroshima/Potsdam.shtml
28. T.N. Dupuy, *The Evolution of Weapons and Warfare* (Jane's, London, 1982), p. 267
29. G. Herken, *Brotherhood of the Bomb* (Owl Books, New York, 2003), p. 139
30. A.C. Clarke, *Astounding Days* (Gollancz, London, 1990), p. 40
31. H.S. Truman, Statement by the President Announcing the Use of the A-Bomb at Hiroshima (6 August 1945), *Harry S. Truman Library and Museum*, https://www.trumanlibrary.org/publicpapers/index.php?pid=100
32. J.W. Campbell, *Astounding Science Fiction* (November 1945), p. 6
33. J.W. Campbell, *Astounding Science Fiction* (December 1945), p. 100

34. J. Horgan, Bethe, Teller, Trinity and the End of Earth, *Scientific American* (August 2015), https://blogs.scientificamerican.com/cross-check/bethe-teller-trinity-and-the-end-of-earth/
35. G. Orwell, You and the Atomic Bomb, Tribune (19 October 1945), http://orwell.ru/library/articles/ABomb/english/e_abomb
36. D. Aaronovitch, 1984: George Orwell's road to dystopia, in *BBC News* (February 2013), http://www.bbc.co.uk/news/magazine-21337504
37. G. Orwell, *Nineteen Eighty-Four* (online text), http://orwell.ru/library/novels/1984/english/
38. J.W. Campbell, *Astounding Science Fiction* (January 1946), p. 6
39. G. Herken, *Brotherhood of the Bomb* (Owl Books, New York, 2003), p. 66
40. G. Herken, *Brotherhood of the Bomb* (Owl Books, New York, 2003), p. 154
41. D. Baker, *Nuclear Weapons* (Haynes, Yeovil, 2017), p. 75
42. G. Herken, *Brotherhood of the Bomb* (Owl Books, New York, 2003), p. 236
43. G. Harry Stine, *ICBM: The Making of the Weapon that Changed the World* (Orion Books, New York, 1991), pp. 162, 163
44. A.C. Clarke, *2010: Odyssey Two* (Granada, London, 1982), p. xiii
45. S.D. Tucker, *Space Oddities* (Amberley, Stroud, 2017), p. 146
46. J. Gifford, The Nature of Federal Service in Robert A. Heinlein's Starship Troopers, https://www.nitrosyncretic.com/pdfs/nature_of_fedsvc_1996.pdf
47. R.A. Heinlein, *Starship Troopers* (Hodder, London, 2015), pp. 140, 141
48. A. Boucher, *Herald Tribune* review of *Starship Troopers*, http://www.panshin.com/critics/PITFCS/133boucher.html
49. R.A. Heinlein, *Starship Troopers* (Hodder, London, 2015), p. 65
50. *Revision of the Commandant's Professional Reading List*, United States Marine Corps, May 2017, http://www.marines.mil/News/Messages/Messages-Display/Article/1184470/revision-of-the-commandants-professional-reading-list/
51. D. Baker, *Nuclear Weapons* (Haynes, Yeovil, 2017), pp. 77, 78
52. G. Harry Stine, *ICBM: The Making of the Weapon that Changed the World* (Orion Books, New York, 1991), p. 262
53. B. Clegg, *Armageddon Science* (St. Martin's Press, New York, 2010), pp. 250–252
54. G.O. Smith, The Undamned, in *Astounding Science Fiction* (January 1947), pp. 72–98
55. I. Asimov, Breeds there a Man? in *Nightfall 1* (Panther, London, 1971), pp. 104–139
56. P. Ware, *Cold War Operations Manual* (Yeovil, Haynes, 2016), p. 28
57. A.C. Clarke, *Astounding Days* (Gollancz, London, 1990), p. 181
58. I. Asimov, The Gentle Vultures, in *Nine Tomorrows* (Pan, London, 1966), pp. 133–151

Journey into Space

In which science fiction gives the world the idea of space travel—and the world continues to dismiss it as "science fiction" even after liquid-fuelled rockets make it a practical possibility. Coming of age at the very outset of the Cold War, such rockets were snapped up by both sides for use as military weapons—but only one side took the space angle seriously. With the launch of Sputnik 1, the Soviets won a race the Americans barely knew existed. Determined not to make the same mistake twice, they went on to win the next race—landing a man on the Moon. That's as far as it went, though. From that point on, space action moved back to its place of origin—science fiction.

Rocket Scientists

Space travel is the archetypal subject-matter of science fiction. At one time, that's all it was, relying on literally fictitious science to launch objects into space. Jules Verne's super-gun in *From the Earth to the Moon* (1865) is a good example. A gun is something that is physically possible—it obeys the laws of physics—but it would be almost impossible to construct one powerful enough to achieve escape velocity. Even then, the enormous initial acceleration would pulverize any human passengers.

Other writers employed even more fanciful methods to get their characters into space. Perhaps the best-known example is the miraculous gravity-screening "Cavorite" that H. G. Wells invented for *The First Men in the Moon* (1901).

© Springer International Publishing AG, part of Springer Nature 2018
A. May, *Rockets and Ray Guns: The Sci-Fi Science of the Cold War*, Science and Fiction,
https://doi.org/10.1007/978-3-319-89830-8_2

The problem is, such materials simply don't exist—and couldn't possibly exist within the framework of the known laws of physics.

Liquid-fuelled rockets are a different matter altogether. First constructed early in the 20th century, they changed things completely. The world finally had a mechanism—one that was practically possible and consistent with the laws of physics—that could, in principle, launch people into space.

The basics of rocket science—the production of forward momentum via the equal and opposite momentum of the rocket's exhaust gases—had been understood ever since the 17th century, when Isaac Newton formulated his third law of motion. Simple gunpowder-powered rockets went back even further than that, having been around since the middle ages. The Chinese used them in battle as long ago as 1232. Well aware of this type of rocket, Jules Verne featured them in his novel *Around the Moon* (1870)—but only as a way to make small changes in the motion of a projectile after it was already in space.

Apparently, Verne considered it too far-fetched to suggest that a rocket might be capable of launching an object all the way from the Earth's surface into space. In the context of a gunpowder-type rocket, he was absolutely right—the exhaust velocity in this case is simply too slow to produce a significant amount of momentum.

It's a different matter when it comes to the more powerful liquid fuels—such as gasoline and kerosene—that emerged towards the end of the 19th century, initially in the context of internal combustion engines. If the same fuels were applied to a rocket motor, the result—in theory at least—really would be capable of getting into space.

This was something the original proponents of liquid fuelled rockets—people like Konstantin Tsiolkovsky in Russia and Hermann Oberth in Germany—realized right from the start. Newton's third law works just as well in outer space as it does on Earth. And unlike an internal combustion engine, which needs to pull oxygen in from the atmosphere, a rocket can carry its oxidizer along with it—making it independent of its environment.

Before long, rockets had become almost synonymous with the idea of practical space travel. Yet space travel, as it always had been, remained synonymous with science fiction. That created a three-way equation, "rockets = space travel = science fiction", that persisted throughout the first half of the 20th century. For many people—both among the general public and engineering professionals—that implied that rockets weren't something to be taken seriously. For others, however, it meant quite the opposite.

The first of the great rocket pioneers, Konstantin Tsiolkovsky, was an occasional writer of science fiction himself. Not surprisingly, he combined the two interests—and his 1920 novel *Beyond the Planet Earth* was one of the first SF works to feature interplanetary rocket travel. It was meant to be more than just entertainment, too. Tsiolkovsky considered SF to be an ideal way to inspire young scientists and engineers. As he wrote in 1935:

Science fiction stories on interplanetary travel carry new ideas to the masses. They excite interest and bring into being people who sympathize with, and in the future engage in, work on grand engineering and technical rocketry projects. [1]

As far as Tsiolkovsky himself was concerned, he found that writing SF also acted as a useful catalyst for his scientific research:

Many times I assayed the scientific concept through the task of writing space novels, but then would wind up becoming involved in exact computations and switching to serious work. [1]

Alongside Tsiolkovsky, another of the early pioneers of rocketry was Hermann Oberth. He was born in what is now Romania, but at the time—1894—was part of Hungary.[1] Fascinated since childhood by the idea of space travel, Oberth moved to Germany at the age of 25 in order to pursue a career in rocketry. Sadly, however, his doctoral dissertation on the subject was rejected as being too fanciful. Instead he had to pursue his interest in an amateur capacity, joining the recently formed *Verein für Raumschiffahrt*—"Society for Spaceship Travel"—in Berlin.

Together with other members of the society, Oberth served as a technical consultant for Fritz Lang's 1928 film *Die Frau im Mond* ("The Woman in the Moon"), which included the first cinematic portrayal of a space rocket. When the Nazis came to power a few years later, they decided the film gave away too many militarily useful secrets. In the words of Fritz Lang:

The orbits around the Earth, and to the Moon and back, were so accurate that the Gestapo confiscated not only all models of the spaceship but also all foreign prints of the picture. [2]

By the 1930s, rockets had become the most instantly recognizable icon of science fiction—especially in the pulp magazines, where SF was just taking off as an independent genre (see Fig. 1).

---

[1] This makes Oberth the fourth Hungarian-born scientist to make an appearance in this book. Leo Szilárd, Edward Teller and John von Neumann all appeared in the first chapter—and the latter two will feature again in later chapters.

Fig. 1 Early SF magazines—such as this issue of *Astounding Stories* from April 1936—often carried surprisingly realistic depictions of space rockets long before such things existed (public domain image)

The situation was encapsulated in a non-fiction article printed by *Astounding Stories* in March 1937:

> The case of the rocketship is a peculiar one. It does not yet exist, but many thousands of people feel certain that there will be a day when the first rocketship leaves the Earth in an attempt to reach interplanetary space, later to soar to the Moon and to the other planets. Thus we have—with the possible exception of flying machines—for the first time in history a machine that is popular even before its actual invention. . ... They occur in the movies; they are comparatively abundant in the daily and periodical press and on the radio, and they are an every-issue event in a good many magazines. [3]

The author of that article, Willy Ley, was a recent arrival in America, having emigrated from Germany two years earlier. He'd been an active member of the *Verein für Raumschiffahrt*, and had worked alongside Oberth on *Die Frau im Mond*. Over the next decade or so Ley worked as hard as anyone to communicate the feasibility and potential of space rockets to the American public.

On the other side of the Atlantic—and working just as hard to deliver the same message—was the British Interplanetary Society (BIS). The link between rockets and science fiction—or rather, between an interest in the two subjects—was just as strong in Britain as it was in America, as Nigel Watson makes clear in this account of the origin of the BIS:

> The germ of its formation was publicized in the January 1933 issue of *Chambers Journal*, by Philip E. Cleator. Throughout the year, Cleator worked to establish a British organization that would popularize the concept of space travel and conduct its own research, much like existing rocket groups in the US, the Soviet Union and Germany. . ... Like Cleator, many of the first members were young, intelligent, avid science fiction fans. Cleator worked hard to promote the group and established the *Journal of the BIS* in 1934, now the longest running astronautical publication in the world. [4]

In those days, science fiction was very much a US-centric genre, even among British readers. For that reason, BIS secretary Leslie Johnson chose to publicize its existence in a letter—aimed at British readers—that was printed in the American magazine *Amazing Stories* in April 1934:

> Its objects are the stimulation of public interest in the possibilities of interplanetary travel and the dissemination of knowledge concerning the problems which at present hinder the achievement of interplanetary travel. [5]

Johnson's ploy worked, and his letter was seen by a number of British SF fans who had previously been unaware of the BIS's existence. Among them was Eric Frank Russell—like Cleator and Johnson, a resident of Liverpool—who went on to become a key member of the BIS, and eventually a popular SF author in his own right.

In due course, the BIS boasted an even more famous member of the SF community: Arthur C. Clarke, who served as the society's treasurer in 1937, and later as its president. He stressed, however, that the activities of the BIS in those days were constrained—by circumstances beyond their control—to be largely theoretical. As he put it: "the actual building of rockets was frowned upon, for it would only result in police proceedings under the 1875 Explosives Act" [2].

The authorities in America took a more lenient attitude, and rocket experimentation had been going on there since the pioneering work of Robert Goddard in the 1920s. Another step forward occurred in 1936, when the California Institute of Technology—or Caltech for short—set up its Rocket Research Project, under the overall direction of Theodore von Kármán (another Hungarian, incidentally). This meant that, for first time, liquid-fuelled rockets were being built and tested in an academic environment.

One member of von Kármán's rocket team was Hsue-Shen Tsien, who has already been mentioned in the previous chapter, "The Super-Bomb". One of America's top rocket scientists, he might have gone on, in the 1950s, to help the country beat the Soviets into space (a race they actually lost, as we will see in due course). Instead, he was deported for having a communist-sounding name.

Also part of the Caltech rocket group was a character named Jack Parsons. As close as science fiction and rocket science always were in the early days, no one could equal Parsons for having a foot in both camps. He was a prominent member of SF fandom, and knew many of the genre's top writers personally. As Colin Bennett wrote in *Fortean Times*:

> Amongst the many areas in which Parsons's influence was felt ... was the burgeoning West Coast science fiction scene. Many key SF writers could be found gathered at the Parsons household in the early 1940s, including Jack Williamson, A. E. van Vogt, Robert Heinlein, Alva Rogers and Forrest J. Ackerman. Ackerman ran the Los Angeles SF society, where Parsons also met Ray Bradbury who professed to being fascinated by "his ideas about the future". [6]

As the same article also points out, Parsons was an unusual individual in other ways too. He was fascinated by pagan and "alternative" religions, and an avowed disciple of the British occultist Aleister Crowley:

> Before each test launch, he was in the habit of invoking Aleister Crowley's hymn to Pan, the wild horned god of fertility. Parsons was an active member of the California Agape Lodge of the sex magical group Ordo Templi Orientis. [6]

In 1943, Caltech's Rocket Research Project acquired a new designation: the Jet Propulsion Laboratory, or JPL. That's a name that's lasted to this day—it's the place that designs and operates most of NASA's interplanetary space probes. Yet no matter how respectable it's become, JPL has never quite shaken off its association with the distinctly unconventional Parsons. In the words of his biographer John Carter:

In the aerospace community, there is a joke that JPL actually stands for "Jack Parsons Laboratory" or even "Jack Parsons Lives". [7]

It may seem odd that an establishment that has been dealing with rockets since its inception has the word "jet" rather than "rocket" in its title. Ultimately, the name is a euphemism, as Arthur C. Clarke explains: "it was called the Jet Propulsion Lab because, in those days, rockets could not be mentioned in polite scientific society" [8].

The broader scientific community's scepticism towards the subject of rockets is something that Clarke—as a member of the BIS—had first-hand knowledge of. As he wrote in his "science-fictional autobiography" *Astounding Days*:

Looking back on it, I am amazed at the half-baked logic that was used to attack the idea of spaceflight; even scientists who should have known better employed completely fallacious arguments to dispose of us. They were so certain that we were talking nonsense that they couldn't be bothered to waste sound criticism on our ideas. [9]

In another book, *Profiles of the Future* (1962), Clarke describes the outrage of the establishment when BIS founder Philip Cleator wrote a book called *Rockets through Space* in 1935. Reviewing the book for the prestigious scientific journal *Nature*, the astronomer Richard Woolley[2] wrote:

It must be said at once that the whole procedure sketched in the present volume presents difficulties of so fundamental a nature that we are forced to dismiss the notion as essentially impractical. [10]

Woolley was wrong. The problems of rocket science aren't "fundamental"—they don't violate any laws of physics—they're just very difficult. So are many other engineering problems, and they can usually be solved by throwing sufficient resources at them. The world being the way it is, that's far more likely to happen if the end result has a military as well as a purely scientific benefit. That was true of rockets—but for a long time, no one realized that. Except the Germans.

---

[2] Woolley's fame today rests almost entirely on his dogmatic pronouncement "space travel is utter bilge", made in 1956, a year before Sputnik 1 [10].

Traditional artillery works in the same way as Jules Verne's space gun—by accelerating a shell for as long as it takes to reach the end of the barrel. After that, the shell follows an unpowered path—technically known as a ballistic trajectory. A rocket, on the other hand, continues to be accelerated for as long as its motor is running. If it runs long enough, and accelerates fast enough, it can achieve Earth orbit—something no real-world gun could ever do. Even if it doesn't make it into orbit, however, a rocket can still vastly outperform a gun-launched shell. When its motor eventually burns out, it simply continues on its way, on an unpowered ballistic trajectory, until it comes back down to Earth at some distant point.

A rocket that makes it all the way into orbit is a space launcher. A rocket that doesn't is something else, and something that is potentially much more deadly—a ballistic missile.

This idea appealed to the Germans for a very good reason. As the losing side in World War One, the country was prohibited by the Treaty of Versailles from possessing most types of military weapon, including large guns and air-dropped bombs. However, the treaty said nothing about rockets.

The military potential of rockets was brought to the attention of the German army by one of the younger members of the *Verein für Raumschiffahrt*. His name was Wernher von Braun, and 30 years later he would design the rocket that would take American astronauts to the Moon. In 1937, however, his priorities were rather different, and he moved to Peenemünde on the Baltic coast to take on technical leadership of the Army Research Centre there. Within a few years, von Braun's A-4 rocket had evolved into the world's first operational ballistic missile—the V-2 (see Fig. 2).

With Europe embroiled in World War Two, Germany's enemies should have been deeply worried about the V-2. Instead, they glibly dismissed it as a propaganda trick. To quote Arthur C. Clarke on the subject:

> With a take-off weight of 12 tons and a range of 200 miles, it was far in advance of anything else that existed at the time; so much so, in fact, that many of Prime Minister Churchill's wartime advisors refused to believe in its existence. [11]

This was in spite of good quality data on the V-2 from a network of spies. As historian Thom Burnett explains:

> The intelligence was not believed. Why? The principle reason was that the British Admiralty thought it was too good to be true and therefore had to be a devious plant by the Abwehr, the German intelligence service. The fantastical claims were written by psychological warfare experts to scare the British. [12]

Fig. 2 The first A4 test rocket on the launch stand at Peenemünde in March 1942 (German Federal Archive, CC-BY-SA 3.0)

One of the few people in the British scientific establishment to take the V-2 seriously was Professor R. V. Jones, who tried in vain to convince his peers of the threat it posed:

> The naïveté of our "experts" was incredible. . ... I can remember a fellow of the Royal Society saying that he was amazed at the accuracy with which the Germans would have to set the rocket before launching it. . ... We knew that the Germans were using gyroscopic control, with information transmitted to rudders in the main jet. As we explained the system, our scientist looked heavenwards and said, "Ah, yes, gyroscopes! I hadn't thought of them!" And that was about the level of the better contributions from the experts. [11]

The Secret Weapon

Although the first A4 rocket was test-fired at Peenemünde as early as March 1942, it was another two and a half years before its weaponized variant—the V-2—entered operational service. The delay was largely due to a lack of enthusiasm on the part of the Nazi leader, Adolf Hitler—for a reason that is possibly even more "sci-fi" than the idea of the rocket itself. To quote Thom Burnett:

> The real reason Hitler was unhappy with the rocket was because of a dream. . . . Hitler believed in the Vril, an intelligent energy force that surrounded and protected the planet, and his dream had shown V-2s igniting the upper atmosphere and destroying this protective force field. [13]

Eventually, however, Hitler was won over. The American rocket scientist G. Harry Stine—who also wrote SF under the pen-name Lee Correy—explains how it happened:

> Von Braun showed a colour motion picture of the first successful A4 launch and how the missile would be used in the field. For the first time, Hitler showed real interest. . . . Hitler said, "I thank you. Why was it I could not believe in the success of your work? If we had had these rockets in 1939 we should never have had this war." [14]

Unfortunately for the Nazis—and fortunately for the rest of the world—the V-2 only became operational after the Allied invasion of Europe in 1944, by which time Germany had all but lost the war. The first rockets were launched against Paris and London on 8 September 1944—and in the course of the next 6 months, over 3000 more were fired, most of them against London and the Belgian city of Antwerp.

Because the missiles were effectively unguided, many of them fell in open countryside rather than hitting the target city. However, the ones that did reach heavily populated areas caused a lot of damage. In London, for example, there were close to 3000 fatalities, and whole streets were reduced to ruins (see Fig. 3).

Needless to say, the devastation caused by the V-2 would have been far worse if it had been fitted with a nuclear warhead. That was a real possibility, because the Germans understood the principle of the atomic bomb as well as anyone. As President Truman said in his speech immediately after the Hiroshima bomb:

Fig. 3 A Londoner inspects the remains of a V-2 rocket motor, with the devastation caused by the rocket visible in the background (public domain image)

We knew that the Germans were working feverishly to find a way to add atomic energy to the other engines of war with which they hoped to enslave the world. But they failed. We may be grateful to Providence that the Germans got the . . . V-2s late and in limited quantities and even more grateful that they did not get the atomic bomb at all. [15]

The V-2 wasn't the only operational liquid-fuelled rocket of World War Two. The Germans also produced a high-speed interceptor aircraft, the Messerschmitt Me 163 Komet, powered by a Peenemunde-designed rocket motor. With a maximum speed close to 1000 km/h and a rate of climb of more than 10 km per minute, it entered service in June 1944, a few months before the V-2. The Komet was quite unlike any other aircraft of its time, as this entry from the *Encyclopedia of Military Aircraft* makes clear:

Maximum powered endurance was eight minutes. With its fuel exhausted the Me 163 would make high-speed gliding attacks on its targets, using its two MK108 30mm cannon and Revi 16B gunsight. With its 120 rounds of ammunition used up and its speed beginning to drop, the Komet would then dive steeply away from the combat area and glide back to base, landing on a skid. This in itself was a hazardous procedure, as there was always a risk of explosion if any unburnt rocket fuel remained in the aircraft's tanks. Many Me 163s were lost in landing accidents. [16]

Slightly less bizarre than the Komet were the first-generation jet fighters, which also entered service—on both sides—during the course of 1944. The Germans had the Messerschmitt Me 262, the British the Gloster Meteor and the Americans the Lockheed P-80 Shooting Star. Like a rocket, a jet engine produces thrust via the momentum of its exhaust—but with the difference that most of the exhaust mass is drawn in from the atmosphere, rather than being carried along as rocket propellant.

As with the atom bomb, the coming of the jet aircraft led to an "I told you so" response from the SF community. Here is John W. Campbell, writing in *Astounding Science Fiction* in April 1944:

Since the Army announced the jet-propelled plane, many a science fiction author, reader—and editor—has discovered that friends, neighbours and acquaintances are abruptly beginning to believe that rocketships aren't exclusively the province of wild fantasy, screwball inventors and impractical dreamers. [17]

Campbell then goes on to point out—quite correctly—that from a pure physics point of view, there is really very little difference between a jet and a rocket:

The jet engine is, of course, a modified rocket engine, and operates on the same essential principles. The main difference lies in the ratio between fuel mass, expelled mass, and fuel-energy-to-expelled-mass. The true rocket expels its fuel, so that fuel mass and expelled mass are equal ... For the jet-propulsion plane ... the mass to be expelled is picked up at the front of the plane, the fuel simply furnishes energy. [17]

For writers and readers of science fiction, the emergence of practical jets and rockets during World War Two meant that space travel must surely be just around the corner. Not everyone saw things that way, however. The V-2 had been a weapon, not a spaceship—and if it had anything to say about the future, it was to point the way towards bigger and better weapons. Since the Soviet Union and the United States were just embarking on the Cold War, that was as important as ever. In the words of Harry Stine:

The Soviets knew what the Americans knew. The ultimate weapon, the Inter-continental Ballistic Missile (ICBM), was now possible. It would be created by marrying the ballistic missile with the atomic bomb. True, the ballistic missile was represented then only by the small V-2, with its range of 250 km. And the atomic bombs of the time weighed more than 10,000 pounds, which was more than the V-2 could carry. But with some concentrated engineering, visionary experts on both sides of the Atlantic Ocean knew the ICBM was possible. [18]

Virtually every vertically-launched, liquid-fuelled rocket of the Cold War—whether a ballistic missile or a space launcher—can trace its heritage in one way or another to the V-2. After the defeat of Germany, unused V-2s were captured by both the Soviets and the Americans. The latter acquired an even greater prize, in the form of Werner von Braun and many of the other Peenemünde scientists. They became naturalized US citizens, transferring their allegiance from the German army to the American one.

The Soviets took a different approach. Instead of relying on German scien-tists, they put their own top rocket man, Sergei Korolev, on the case. After finding out how the V-2 worked—by taking captured examples apart—he began designing new rockets of his own. The result, in many cases, was better than anything his German-American rivals were doing at the same time.

Although most of the rockets produced in the 1950s—both in America and Russia—owed a debt to the V-2 in terms of their internal workings, they looked nothing like it externally. The V-2 had a distinctive rounded shape, with prominent tail fins. On the other hand, American rockets tended to be long, thin cylinders with much smaller tail fins, while Korolev in Russia went for a tapering cluster of rocket motors.

That was how things were in the real world, anyhow. In the science fiction of the 1950s, almost every depiction of a rocket continued to look more like a V-2 than anything else (see Fig. 4).

The Race into Orbit

The basic theory of orbits, like that of rockets, can be traced back to Isaac Newton, and to a specific book he wrote in 1687: *Philosophiae Naturalis Principia Mathematica*, or "Mathematical Principles of Natural Philosophy". In addition to his three laws of motion, Newton describes an idealized "thought experiment" involving a cannonball fired from the top of a high mountain (see Fig. 5).

January 1956 · 35 Cents

Astounding
SCIENCE FICTION

Fig. 4 Throughout the 1950s, depictions of space rockets in popular culture almost always bore a close resemblance to the V-2, as in this example from the January 1956 issue of *Astounding Science Fiction* (public domain image)

Ignoring the effects of air resistance (the main reason this idealized scenario wouldn't actually work in the real world), the only force acting on the cannonball after it leaves the barrel of the gun is gravity. This, of course, pulls it towards the centre of the Earth. However, there's another of Newton's laws to think about: his first law of motion, which states that once in motion an object has a natural reluctance—technically called "inertia"—to alter its motion.

The cannonball's inertia means that, as its launch speed is increased, it travels further and further before gravity pulls it back to Earth. Since the latter is a sphere, "travelling further and further" actually means "curving further and further around the planet". If the cannonball's speed is high enough, it will travel all the way round without ever falling back to Earth. In other words, Newton's cannonball will be in orbit.

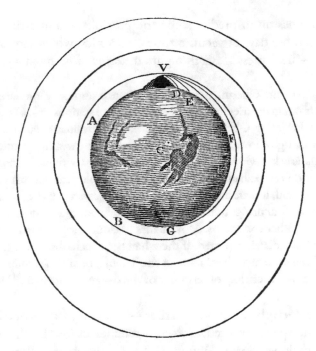

Fig. 5 An illustration from an 1846 edition of Isaac Newton's *Principia*, showing how a cannonball launched from a high mountain V travels increasing distances to points D, E, F and G as its speed is increased. At even higher speeds, the cannonball goes into orbit around the Earth (public domain image)

Although this is 17th century physics, it still seemed like science fiction to most people in the early 20th century. Ironically, the SF community itself was among the first to recognize that space travel might soon be fact rather than fiction. This was a point that was made by Hugo Gernsback in the April 1930 issue of *Air Wonder Stories*, in what was effectively an updated version of Newton's thought experiment:

> It will be necessary only to build a rocketship and elevate it beyond the appreciable atmosphere of the Earth—say a trifle over 500 miles—then give it a sufficient impulse in a direction at right angles to the position of the Earth. It will then continue to gravitate around the Earth without falling; thus becoming a new satellite; and it will maintain its orbit permanently until it is disturbed by some external force. Of course, at such a distance, it is to be supposed that no atmospheric friction will be encountered to reduce the original speed—which must be in the order of five miles a second. [19]

Gernsback goes on to predict some the possible uses of artificial satellites, although—in those days of vacuum tube electronics, which required constant maintenance—he expressed his thoughts in terms of manned space stations:

> Professor Hermann Oberth, perhaps the greatest authority on interplanetary space, points out many uses for such revolving "space stations". . . . One important purpose, as Professor Oberth points out, is the invaluable aid that such an observatory can give to the science of meteorology, or weather prediction, as it is more popularly known. If the observatory is equipped with radio, instantaneous communication can be had with the various meteorological stations scattered all over the Earth and, if there are a number of such observatories circling around the Earth (let us say four or eight), they can immediately notify any station on Earth as to probable weather conditions. Movements of clouds; fog formations; icebergs, etc., can be immediately reported. If there had been such observatories years ago, one could have prevented the sinking of the Titanic, because the ship could have been notified by the circling observatory of the dangers in its path. [19]

The figure Gernsback quoted earlier—"five miles a second", or about 8 km/s—is the speed required to keep a satellite in low Earth orbit, similar to the International Space Station (ISS) today. At an altitude of some 400 km, the ISS takes about 90 minutes to complete an orbit—meaning that it goes all the way around the Earth 16 times a day. Satellites at higher altitude take longer to complete an orbit. Part of the reason for this is obvious—they have further to travel. On top of that, however, the pull of gravity is weaker at higher altitude, so less speed is needed to hold an object in a circular orbit.

This leads to an interesting situation. If a satellite's altitude is high enough, there's a point at which it takes exactly 23 hours, 56 minutes to complete an orbit—the same time the planet takes to rotate once on its axis. If such a satellite is positioned directly above the equator, it will remain permanently fixed over the same spot on the Earth's surface. That's the principle behind geostationary satellites—and it's why a TV satellite dish can always pick up a signal even though it never moves.

The existence of geostationary orbits is a long-established consequence of basic physics. However, the specific use of such orbits for radio and TV broadcasting was first proposed by Arthur C. Clarke—best known as a science fiction author, but on this occasion applying his ingenuity to non-fiction speculation. In recognition of his contribution, the region of space occupied by geostationary satellites is now called "the Clarke Belt".

Clarke's suggestion was originally published in the October 1945 issue of *Wireless World* magazine, in the form of a four-page article entitled "Extraterrestrial Relays". A couple of decades later, in 1966—two years after the first geostationary communication satellite, or comsat, was launched—he wrote a follow-up piece forlornly called "A Short Pre-History of Comsats, or How I Lost a Billion Dollars in My Spare Time":

> It is with somewhat mixed feelings that I can claim to have originated one of the most commercially valuable ideas of the 20th century, and to have sold it for just $40. [20]

When Arthur C. Clarke wrote "Extraterrestrial Relays" in 1945, most people imagined that artificial satellites and space travel were still as far-fetched as ever. In fact, after Germany's success with the V-2, the most important engineering obstacles had already been overcome. The problem was no longer a technical one—"how do we get into space?"—but a philosophical one: "why should we go into space?" That's a point Robert A. Heinlein had one of his characters make in "The Man Who Sold the Moon" (1950):

> The real engineering problems of space travel have been solved since World War Two. Conquering space has long been a matter of money and politics. [21]

By that time, "politics" meant the East-West politics of the Cold War. The situation played out in much the same way as the race for the H-bomb, described in the previous chapter. The Russians quietly got on with it, while the Americans tied themselves in knots arguing about the pros and cons. That's something that bothered another Heinlein character, a retired US Navy admiral, in the short novel he wrote based on the 1950 movie *Destination Moon*:

> The United States is going to stall around and let Russia get to the Moon first—with hydrogen bombs. [22]

And later:

> We'll wake up one morning to find Russia with a base on the Moon and us with none—and World War Three will be over before it starts. [23]

Heinlein clearly understood that the conquest of space had become politically inseparable from the Cold War. Probably most ordinary Americans did too. At an official level, however, it was something the US establishment was reluctant to admit. The result was an artificial compartmentalization, which saw space-related research completely separated from weapon development. The latter—particularly work on ballistic missiles—was given a high priority, while the former, as mere "science", was given a low one.

Wernher von Braun and the other German rocket scientists were put to work on the military side. On behalf of the US Army, they developed a new missile with the rather uninspiring name of Redstone—after the Redstone Arsenal in Huntsville, Alabama, where they were based. The Redstone had a similar range to the V-2, but a significantly larger payload capacity—allowing it to carry a multi-megaton nuclear warhead.

Longer range missiles—the kind that might double up as space launchers—were considered to fall in the domain of the US Air Force rather than the Army (the logic being that they filled a role closer to that of bomber aircraft than artillery guns). As such, it was the Air Force that was tasked with developing America's first ICBMs—defined as having a range of 5500 km or more. It was a project that von Braun, as an Army employee, had nothing to do with.

The Soviet Union adopted a more streamlined approach, with their top rocket scientist, Sergei Korolev, at the helm. He started out with short range missiles—essentially derivatives of the V-2—and then went straight on to the development of Russia's first ICBM, the R-7. In a slightly modified form, this also served as the country's first space launcher (see Fig. 6). It was essentially an R-7 missile that put the first satellite, Sputnik 1, into orbit in 1957—and a later version, with more powerful engines, that launched the first human being, Yuri Gagarin, into space in 1961. A distant descendant of Korolev's R-7 is still used to launch Soyuz spacecraft today.

In contrast, the situation in the United States was increasingly chaotic. All three services—the Army, Navy and Air Force—were developing missiles independently of each other, and many of them were crashing or going astray on test flights. Such incidents were widely publicized—and opportunistically picked up by Ian Fleming for his James Bond novel *Dr No* (1958).[3] The book's eponymous villain, at his Caribbean hideaway of Crab Key, takes credit for many of those real-world failures:

> Perhaps you have read of the rockets that have been going astray recently? The multi-stage Snark, for instance, that ended its flight in the forests of Brazil instead of the depths of the South Atlantic? ... You recall that it refused to obey the telemetered instructions to change its course, even to destroy itself. It developed a

---

[3] Unlike its film adaptation, the book version of *Dr No* was the 6th instalment in the James Bond series, not the first.

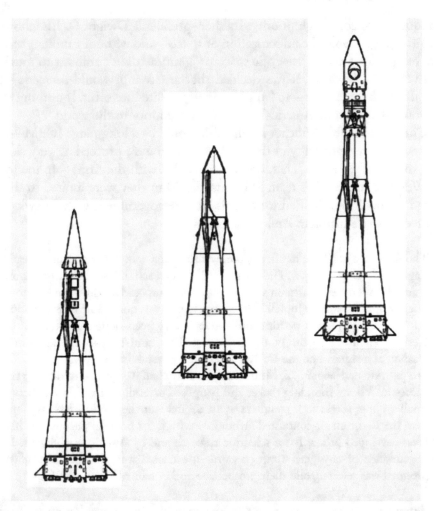

Fig. 6 The launcher that put Sputnik 1 into orbit in 1957 (centre) and the Vostok launcher used for the first manned spaceflights (right) were both variants of the original R-7 ICBM shown on the left (NASA image)

will of is own? ... There have been other failures, decisive failures, from the long list of prototypes—the Zuni, Matador, Petrel, Regulus, BOMARC ... it may interest you to know that the vast majority of those failures have been caused from Crab Key. [24]

All the missiles mentioned by Dr No—which really did crash, though not necessarily for the reason he states—were intended as weapons, not space launchers. Unlike the Soviets, the American administration didn't give the conquest of space a high priority. Its then-president, Dwight D. Eisenhower, was at pains to avoid the militarization of space—and without a military angle, the budget for a purely scientific space programme didn't amount to much.

Eventually, however, it was decided that an attempt would be made—by one of the three services—to launch a small satellite into orbit. If nothing else, it would demonstrate America's technical superiority to the world.

The most suitable launcher for the task would be a long-range ICBM—but that was the responsibility of the Air Force, who were proceeding very slowly due to lack of expertise. That, of course, resided with the Army—in the form of Wernher von Braun's team at Huntsville—but they were limited to short-range weapons like the Redstone. This left the government with an awkward choice, as space historian Paul Drye explains:

> The Air Force had little to offer as their Atlas rocket was still at a relatively early stage of development.... The Army and Navy each made their cases more plausibly. Wernher von Braun's Huntsville team proposed modifying a Redstone nuclear missile into the Jupiter C as a launcher. It was quite close to completion but suffered from two political problems: it had been designed primarily by German engineers, many of them former Nazis, and it was derived from a weapon at a time when the US was interested in establishing space exploration as a peaceful endeavour . . ... Meanwhile the Navy had worked . . . to develop the successful Viking sounding rocket and proposed extending it with another two smaller upper stages so it could act as an orbital launcher. While less far along than the Redstone, it contrasted favourably with it for political purposes: it had been developed primarily for scientific research and by American engineers. In the absence of any great time pressure, the choice was obvious. The Navy proposal was selected, and their prospective rocket named Vanguard. [25]

America's inter-service "space race" was prophetically foreshadowed by the science fiction author William Tenn. In his short story "Project Hush"(1954), the US Army sets up a top secret base on the Moon—only to discover another, larger, base not too far away. After going out to investigate, one of the Army officers, Major Gridley, reports back to the mission commander (the story's first-person narrator):

> "The other dome—you want to know who's in it. You have a right to be curious, Ben. Certainly. The leader of a top-secret expedition like this—Project Hush they call us . . . finds another dome on the Moon. He thinks he's been the first to land on it, so naturally he wants to—"
> "Major Monroe Gridley!" I rapped out. "You will come to attention and deliver your report. Now!" . . .

"They aren't extraterrestrials in the other dome, Ben," Monroe volunteered in a sudden burst of sanity. "No, they're human, all right, and from Earth. Guess where."

"I'll kill you," I warned him. "I swear I'll kill you, Monroe. Where are they from—Russia, China, Argentina?"

He grimaced. "What's so secret about those places? Go on—guess again."

I stared at him long and hard. "The only place else—"

"Sure," he said. "You got it, Colonel. The other dome is owned and operated by the Navy. The goddamn United States Navy!" [26]

In the real world, the situation became just as ludicrous—if not more so. By 1956, von Braun's team—who weren't supposed to be building a space launcher—were ready to test a rocket, called Jupiter-C, that was effectively just that. This was a whole year before the Russians launched Sputnik 1, and the US Army could have beaten them to it. But rules were rules, and the army wasn't allowed to put things into orbit—even by accident. As Thom Burnett recounts:

The first Jupiter-C rocket, numbered RS-27, was launched from Cape Canaveral on September 20th, 1956. It was essentially a Redstone first stage with two extra solid-propellant rockets developed by the Jet Propulsion Lab at Caltech. The Jupiter-C flew 3355 miles, reaching an altitude of 682 miles and a velocity of Mach 18. Because of the Department of Defence's directive that von Braun was not to attempt a space launch, the Jupiter-C carried a dummy payload as its fourth stage. The engine was filled with sand ballast instead of solid fuel in order to prevent an "accidental" boost of the fourth stage into orbit. . . . Von Braun later claimed that Washington had been so fearful the Army Ballistic Missile Agency would go ahead with an unauthorized satellite launch that two observers were sent from the White House Bureau of the Budget to ensure this did not happen. [27]

The Sputnik Crisis

If Wernher von Braun had been allowed to put a fuel load, rather than ballast, in the fourth stage of the Jupiter-C in 1956, the United States would have entered the history books as the winner of the space race. That would, at the time, have come as a huge surprise to the American public—most of whom didn't even realize such a race was on. It's difficult to believe in hindsight, particularly when the subject is approached through the niche lens of the SF community, but space just wasn't something Americans gave much thought to in the 1950s.

Thus it was that history had a surprise in store for them anyway—with the launch, not of an American rocket on 20 September 1956, but a Russian one on 4 October 1957. This was one of Sergei Korolev's R-7s, and it put the world's first artificial satellite into orbit: a 58-centimetre aluminium sphere called Sputnik 1. For a nation that considered itself the technological leader of the world, the launch of Sputnik 1 made a huge dent in America's pride.

It was another two months before the US Navy was finally ready to launch its own satellite, Vanguard. Scheduled for December 6, it was heralded by a huge amount of publicity, and the launch was covered live on television. Indeed, as space launches go, it was a spectacular one. The rocket lifted a metre or so off the ground, then promptly exploded in a huge ball of flame. Having missed the chance to become the first nation in space, America had succeeded in making itself a laughing stock. To quote historian Thom Burnett:

> The world's press had a field day thinking up humorous names to call the pathetic Vanguard. The *New York Times* dubbed it "Sputternik", the *New York Herald Tribune* called it "Goofnik", *Time* magazine referred to the whole programme as "Project Rearguard", the *News Chronicle* labelled it "Stayputnik", the *Daily Express* "Kaputnik". [28]

At long last the US establishment relented, and gave the Army the go-ahead to launch a satellite using one of von Braun's Redstone-derived rockets. Called Explorer 1, it was finally launched on 31 January 1958—not even the same calendar year as Sputnik 1. That was much too late, and Explorer 1 was merely "exploring" where the Soviets had already been.

The loss of the space race had a huge impact on the United States, demolishing its self-image as the world's most technologically advanced nation in the most dramatic way possible. It produced an outbreak of frenzied heart-searching, dubbed the "Sputnik crisis", at all levels of American culture—including, of course, the SF magazines.

In January 1958, the *Magazine of Fantasy and Science Fiction*'s cover feature wasn't fiction at all, but a non-fiction article entitled "Sputnik: One Reason Why We Lost" (see Fig. 7). The article's author, G. Harry Stine, was better known in the SF community as "Lee Correy"—a pseudonym he used for the occasional fictional pieces he produced. In his day job, however, he worked as a missile engineer for the Glenn Martin aircraft company—or he did, until exactly one day after the launch of Sputnik 1. That was the day he was quoted in the press as saying "we have underestimated the Russians all along" [29]—and was promptly fired by

Fig. 7 Russia's success with Sputnik 1 provoked a frenzy of self-criticism in America—such as the article "Sputnik: One Reason Why We Lost" in the January 1958 issue of *The Magazine of Fantasy & Science Fiction* (public domain image)

his employers. If nothing else, this left him free to expand on the same theme in the magazine article.

Although Stine's article runs to six pages, it comes down to a single, very frank assessment of the situation: the Soviets won the space race because they had better scientists and engineers than the Americans.

A similar message can be found in another article, "Behind the Sputniks", in the April 1958 issue of *Fantastic Universe*. The author in this case was Lester del Rey, who had been writing science fiction since the 1930s—around the time Konstantin Tsiolkovsky pointed out the benefits of the genre for inspiring young scientists and engineers. That's pretty much the point del Rey makes in his article, going so far as to link Sputnik's success to Russia's habit of taking SF seriously:

Russia is working harder at the job of producing future scientists and engineers than we are … Once in a great while, we even see that Russia has been using science fiction magazines—of the old-style gadgetry, pro-technology type—to inculcate a love of science into the younger minds and convince others that science is not evil. It's a pity to see our own development turned against us this way, since we used to regard science fiction as something almost as purely American (or at least, Anglo-American) as would be the space-flights about which we dreamed. [30]

In hindsight, the Sputnik crisis wasn't a bad thing. It forced Americans to take the subject of space more seriously, and persuaded them that overtaking Russia in the space race was a national priority. It's possible, in fact, that the perception that America was falling behind in rocket science helped to determine its next choice of president—as Thom Burnett explains:

Future President John F. Kennedy started campaigning on the basis that there was a "missile gap" between Eisenhower's run-down ballistic forces and the Soviets' evidently superior ones. Although this was inaccurate, the popular perception of military inferiority would help launch Kennedy all the way to the White House after the next election. [31]

Next Stop: The Moon

As far as America was concerned, things got worse before they got better. On 12 April 1961—less than three months after Kennedy's inauguration—the Soviet Union launched its first cosmonaut, Yuri Gagarin, into orbit. America's response was less than spectacular. On 5 May, their own first astronaut, Alan Shepard, made a brief suborbital flight—with just five minutes of weightlessness, compared to Gagarin's 80 minutes. While the latter had been hurled into space by a full-fledged ICBM, Shepard had to rely on von Braun's trusty—but much shorter range—Redstone.

America was on the back foot in space, and Kennedy was determined to reverse the situation. On 25 May 1961, still only four months after his inauguration, he made a speech to Congress on the subject of "Urgent National Needs". It contained one of Kennedy's best known sound-bites—possibly the most ambitious, inspiring statement ever made by an American president:

I believe that this nation should commit itself to achieving the goal, before this decade is out, of landing a man on the Moon and returning him safely to the Earth. [32]

Taken on its own, this sounds like pure scientific altruism. In its context, however, it was much more cynically realistic than that. For Kennedy, space exploration—with its proven ability to grab headlines around the world—was as much about Cold War propaganda as anything else:

If we are to win the battle that is now going on around the world between freedom and tyranny, the dramatic achievements in space which occurred in recent weeks should have made clear to us all, as did the Sputnik in 1957, the impact of this adventure on the minds of men everywhere, who are attempting to make a determination of which road they should take. [32]

The fact that space rockets just happened to employ the same technology as ballistic missiles was an added bonus. Whichever side was ahead in one would also be ahead in the other. That would have been true whether Kennedy had made his speech or not. But the speech added something new to the mix: a clearly defined goal, in the form of a manned landing on the Moon.

As far as the SF community was concerned, that was old news. Ever since the days of Jules Verne, the Moon had always been the prime destination in space. More recently, since the beginning of the Cold War, SF had foreseen that the race to the Moon would end up as a two-way one, between East and West. As Robert A. Heinlein had one of his aspiring rocket scientists say in *Rocketship Galileo* (1947):

The United States isn't the only country on the globe. It wouldn't surprise me to hear some morning that the Russians had done it. They've got the technical ability and they seem to be willing to spend money on science. They might do it. I have nothing against the Russians; if they beat me to the Moon, I'll take my hat off to them. [33]

By the end of the 1950s—still several years before Kennedy's announcement—the notion of Russians and Americans competing in a "Race for the Moon" was well established in popular culture—even becoming the title of a comic book (see Fig. 8).

However, when President Kennedy made his Moon speech in 1961, it wasn't only because of comics and sci-fi films. By that time, several serious studies had been carried out that showed a manned lunar mission to be a practical possibility. Those studies weren't known to the public, though—because they had been done behind a cloak of military secrecy.

As we have already seen, America's top rocket scientist—Wernher von Braun—spent the 1950s hidden away in the US Army's Ballistic Missile

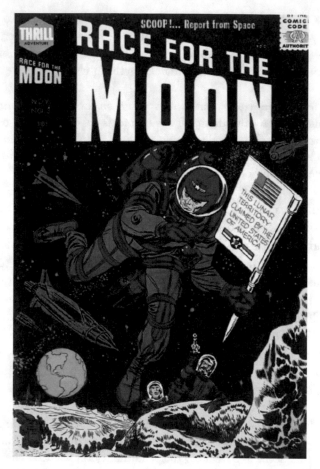

Fig. 8 The November 1958 issue of *Race for the Moon* from Harvey Comics (public domain image)

Agency. His day job was designing short-range tactical missiles like the Redstone and Jupiter. In the background, however, he had bigger ambitions—with plans for a family of much larger rockets called Saturn. These weren't missiles, because the Army was prohibited from working on long-range weapons (that was the job of the Air Force). Instead, the Saturn rockets were space launchers. The Army needed them, so von Braun argued, because it was going to build a base on Moon.

It was called Project Horizon, and the official report, dated 1959, was boldly subtitled "A United States Army Study for the Establishment of a Lunar Military Outpost". This was far more ambitious than the "flags and footprints"

Apollo programme, envisaging a substantial Moon base that would be permanently manned by the late 1960s. To quote from the report itself:

> Initially the outpost will be of sufficient size and contain sufficient equipment to permit the survival and moderate constructive activity of a minimum number of personnel (about 10–20) on a sustained basis. It must be designed for expansion of facilities, resupply, and rotation of personnel to insure maximum extension of sustained occupancy. It should be designed to be self-sufficient for as long as possible without outside support. [34]

In the event, von Braun's proposal failed to secure any funding. As space historian Paul Drye explains:

> Ultimately Project Horizon didn't fool enough people into thinking that the Army knew what they were doing. Even looking past the ... antipathy that President Eisenhower had for the military in space, he was known to have used the words Buck Rogers[4] more than once to describe the nebulous plans he got from the Army and others, and he was justified in saying so. [36]

Just as William Tenn had predicted in his short story "Project Hush", von Braun's Army team weren't the only ones aiming at the Moon. Their great inter-service rivals in the US Air Force came up with their own Lunar Expedition Plan, or Lunex for short. As a secret report from 1961 explains: "the Lunar Expedition has as its objective manned exploration of the Moon with the first manned landing and return in late 1967" [37].

As it turned out, the job of going to the Moon went to neither the Army or the Air Force, but to a civilian government agency: the National Aeronautics and Space Administration, better known as NASA. Wernher von Braun transferred there from the Army in 1960, taking his plans for the Saturn family of rockets with him. The largest of these—the Saturn V—was going to be the launch vehicle for Project Apollo ... and Project Apollo was going to take American astronauts to the Moon.

To start with, Project Apollo was simply a name—with no spacecraft design or mission plan to go with it. Settling on those was NASA's first task after President Kennedy pointed them in the direction of the Moon. Three basic options were considered. The most obvious of these was called "Direct Ascent"—going all the way from the Earth to the Moon and back with a

---

[4] Buck Rogers was a popular sci-fi serial in the days of old-time radio, before TV and blockbuster movies. As SF historian David Kyle put it: "The solemn intonation of the announcer to start each episode, 'Buck Rogers in the 25th century!' followed by rolling thunder, was a familiar sound across the nation." [35]

single vehicle, as depicted in most science-fictional Moon voyages. A variant of this, using several smaller launch vehicles in place of a single large one, was "Earth Orbit Rendezvous", with the lunar spacecraft being put together in Earth orbit rather than on the ground. The third and final option was "Lunar Orbit Rendezvous", in which the main spacecraft remains in lunar orbit while a smaller vehicle makes the landing.

Eventually the last of these was chosen—but the idea wasn't original to NASA. It had been considered by the British Interplanetary Society at least ten years earlier, as Arthur C. Clarke recounts:

> We discussed many types of rendezvous and space-refuelling techniques, to break down the journey into manageable stages. One of those involved the use of a specialized ferry craft to make the actual lunar landing, while the main vehicle remained in orbit. This, of course, is the approach in the Apollo project—and I am a little tired of hearing it described as a new discovery. For that matter, I doubt if we thought of it first; it is more likely that the German or Russian theoreticians had worked it out years before. [38]

The basic design of the Apollo spacecraft had been finalized by the middle of 1962. From then on, it was just a matter of building and testing all the various components and mission phases—a longer and more laborious process than it is usually depicted in SF stories. Nevertheless, NASA still made it to the Moon within Kennedy's "end of the decade" timeframe. Apollo 11, with two astronauts on board, touched down on the lunar surface on 20 July 1969.

At some point in the middle of the 1960s, the Soviets lost their lead in the space race. By this stage the situation had become the mirror-image of the previous decade. In the 1950s, the American space programme had been chaotic and goal-less, while the Russians were single-minded and purposeful. Now it was the other way around. The Soviets no longer seemed to know what they were doing in space, with several unrelated programmes running in parallel, and no clear political backing for any of them.

Part of problem came down to the Soviet Union's communist ideology. Centred almost exclusively on human affairs, communism was inherently Earth-focused. What possible reason could a good communist have for wanting to go to the Moon? From the point of view of Soviet politicians, there was much more useful work to be done in Earth orbit—all the way from the monitoring of industrial and agricultural resources to Cold War espionage.

The Russians' answer to Apollo—the Soyuz spacecraft—kept its options open. Maybe it would take Soviet cosmonauts to the Moon, or maybe it would shuttle them to and from a space station in Earth orbit. The lunar option wasn't ignored altogether, and the Russians went as far as designing a one-man Moon lander called the LK. This would be have been delivered, "lunar orbit rendezvous" style, by a Soyuz spacecraft in the same way that Apollo delivered NASA's lunar module. Compared to the latter, however, the LK looked as second-rate as America's Explorer 1 had been to Sputnik 1—or Alan Shepard's flight to Yuri Gagarin's (see Fig. 9).

Paul Drye describes the Soviet Moon plan in the following way:

> It would have sent two men to the Moon aboard a customized Soyuz, one of whom would then enter the purpose-built LK lunar lander and descend to the surface. Apart from the smaller crew, it was similar in many ways to the Apollo approach. [39]

As he then goes on to point out, "there were a number of places where the Soviet Union made time-wasting mistakes as compared to Apollo" [39]—and the LK never got off the ground. Nevertheless, the fear that it might have done—and that the Soviets might beat America to the Moon—was a spur that gave added urgency to the Apollo programme. Its second manned flight, Apollo 8 in December 1968, was originally scheduled as a simple test in Earth orbit. In the mistaken belief that the Russians were further ahead than they actually were, NASA then switched it to a circumlunar flight. As Apollo 8's commander, Frank Borman, explained:

> I was told that the CIA had intelligence that the Russians were going to try and put a man around the Moon before the end of the year, and that was the reason for the change in our mission. Because, after all, the Apollo programme was just a battle in the Cold War. [40]

A "battle in the Cold War" it might have been, but it was a very low-key battle—with little of the drama that SF authors had expected the first voyage to the Moon would entail. The Apollo astronauts didn't encounter an underground lunar civilization, as H. G. Wells had predicted in *The First Men In The Moon* (1901). They didn't cause riots and public outrage of the kind Isaac Asimov depicted in his 1939 short story "Trends"—or find diamonds lying around, like the space pilot in Robert A. Heinlein's "The Man Who Sold the Moon" (1951).

Nor was the actual Moon race anything like as dramatic as the one portrayed in Jeff Sutton's novel *First on the Moon*—published in 1958, not long after Sputnik 1. The day before the first American Moon rocket is due to blast off, an attempt is made to assassinate its commander. Then, just after the

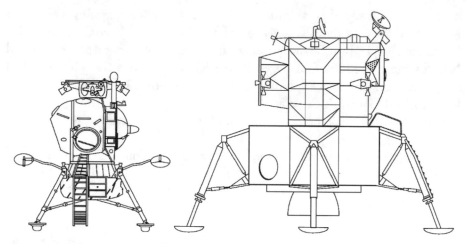

Fig. 9   The proposed Soviet LK lander (left) and the Apollo lunar module (right), drawn to the same scale (NASA image)

launch, a missile is fired at it in a bid to shoot it down. As if that wasn't enough, it's discovered there's a saboteur on board—and a Soviet spacecraft is following close behind. The book's cover blurb isn't exaggerating when it describes it as "a novel of the life-or-death race between the USA and the USSR" [41].

As over-the-top as *First on the Moon* is, Jeff Sutton's next novel about a lunar voyage—*Apollo at Go* (1963)—was as prophetic of real-world events as SF gets. As the title suggests, it was written after the Apollo mission had been defined on paper—but still long before any hardware had been built, and Sutton's research on the subject is immaculate. Just like the real-world Apollo 11, his Moon mission starts from Launch Complex 39 at Cape Canaveral—albeit on Saturday 5 July 1969, rather than Wednesday 16 July.

Here is Sutton's fictional TV commentator describing the launch:

It's a slow climb, deceptively slow, like a fly scaling the side of a skyscraper. The five giant F-1 engines are pouring out a savage torrent of flame—seven and a half million pounds of thrust. Saturn's commencing to pick up speed, clawing at the sky. Within a few seconds she'll begin pitching over, into a trajectory that'll take her into orbit on the first leg to the Moon. [42]

"Deceptively slow" isn't the way many writers of fiction would have envisaged a space launch—but it's a perfect description in the case of a Saturn V. Sutton even gets his figures correct: the enormous first stage of the Saturn V does indeed generate a thrust of "seven and a half million pounds", or about 35 meganewtons.

The Moon landing itself takes place, in the novel, on Tuesday 8 July 1969. Here's part of the tense three-way dialogue between mission commander Joe Faulk (speaking first), lunar module pilot Max Kovac and Earthside capsule communicator (capcom) Whitey Burke:

"Still can't discern detail except the dark area looks more and more like a crevice. I'd say it's pretty wide, six or eight feet. . ... I'm below 300 feet and still guessing. Looks bad."

"Better pick a spot," Burke said edgily.

"Trying, Whitey."

"Two hundred feet," Kovac interrupted. "Fuel 3 per cent ..."

"Getting low on fuel," Faulk broke in. "Can't tell much about area around patch except that surface is smooth. . ... Got a nasty suspicion."

"Ash?"

"Yeah, could be deep."

"Steer away from that, Joe."

"Where? No choice. Big rocks at the border. Still moving down."

"One hundred feet," Kovac called. [43]

Compare that with the transcript of the actual landing, on 20 July 1969. The voices here belong to mission commander Neil Armstrong and lunar module pilot Buzz Aldrin:

Aldrin: 300 feet [altitude], down three and a half [feet per second], 47 [feet per second] forward. Slow it up. One and a half down. Ease her down. 270.

Armstrong: Okay, how's the fuel?

Aldrin: Eight percent.

Armstrong: Okay. Here's a—looks like a good area here.

Aldrin: I got the shadow out there. . ... Three and a half down, 220 feet, 13 forward. . ... Coming down nicely.

Armstrong: Gonna be right over that crater.

Aldrin: 200 feet, four and a half down.

Armstrong: I got a good spot. [44]

As soon as the lunar module is safely down—in the novel, that is—Faulk responds to a query from the third crew member, Les Mallon, who remained in orbit in the main Apollo command module (the counterpart of Mike Collins in the real world):

Faulk became aware of Mallon's worried call on the radio and acknowledged, adding "Bug landed at T plus 64 minutes from start of go-down."

"Congratulations, Skipper. You had us sweating." Mallon's voice displayed relief. [45]

Compare that with the real-world exchange between Neil Armstrong and capcom Charlie Duke:

Armstrong: Engine arm is off. Houston, Tranquillity Base here. The Eagle has landed.

Capcom: Roger, Tranquillity. We copy you on the ground. You got a bunch of guys about to turn blue. We're breathing again. Thanks a lot. [44]

For that real landing—six years after Jeff Sutton wrote *Apollo at Go*—science fiction writers found themselves in high demand as TV pundits. As *Variety* magazine wrote in its preview of the TV coverage before the big event:

A unique aspect is the wedding of science fiction with science fact, as CBS and ABC in particular line up leading sci-fi scribes as guest experts. On CBS, Arthur C. Clarke (author of *2001: A Space Odyssey*) will serve as a special consultant in the same studio as Walter Cronkite, and Orson (*War of the Worlds*) Welles will appear live from London. Welles will narrate a science fiction film prepared by CBS News. On ABC, clips from the famous sci-fi flicks will be unspooled, together with commentary by *Life* movie critic Richard Schickel. ABC's coverage will also feature a panel of sci-fi writers moderated by Rod Serling. Panellists are Isaac Asimov, Frederik Pohl, and John R. Pierce, who writes under the name of J. J. Coupling. [46]

Modern readers may be puzzled by that last name, which isn't as familiar as it used to be. Nevertheless, we will see quite a bit of John R. Pierce—alias J. J. Coupling—in the next chapter, "Electronic Brains".

Lost in Space

Following the excitement of the Moon race in the 1960s, the 1970s can only be described as a disappointment—at least as far as space exploration was concerned. Perhaps the most memorable moment occurred on 17 July 1975, when an Apollo spacecraft docked with a Soyuz, and the Soviet and American crews shook hands (see Fig. 10).

Fig. 10   An artist's conception of an Apollo spacecraft (left) docking with a Russian Soyuz (right), the symbolic high-point of the Apollo-Soyuz Test Project in July 1975 (NASA image)

While the Apollo-Soyuz Test Project may have been a major diplomatic achievement, it was nothing special from a technical point of view. Almost a decade earlier, in his 1968 novel *2001: A Space Odyssey*, Arthur C. Clarke had foreseen Russians and Americans working together in space—but in a much more ambitious context than anything that happened in reality. By the titular date of 2001, Clarke portrays the two countries jointly operating a far more sophisticated space station than the real-world ISS. He also envisaged them having permanent bases on the Moon, and undertaking exploratory voyages to the other planets of Solar System.

To a seasoned reader of science fiction, there was nothing far-fetched about Clarke's vision. It had been a common expectation—since the 1930s, if not earlier—that landing on the Moon would simply be the first step to the planets. No SF story predating 1970 shows humans reaching the Moon, deciding that was further than they needed to go, and retreating back to Earth orbit. Yet that's what happened in the real world.

In an alternate version of history, diverging from real events after Apollo 11's Moon landing in 1969, NASA might have had astronauts on Mars by the middle of the 1980s. They had the technical capability to do that, and detailed plans for such a mission. Just such an alternate timeline is depicted in Stephen Baxter's novel *Voyage* (1996)—which can't really be called science fiction, because there's no fictional science in it. Every item of technology depicted in the novel really did exist in the 1980s, or was under development when the timelines diverged in 1969. The differences are purely political ones, as Baxter explains in his afterword:

> To space proponents in 1969, technical logic appeared to indicate a building from the achievements of Apollo to a progressive colonization of the Solar System, including missions to Mars. But the political logic differed. [47]

Inevitably for those times, "political logic" meant "Cold War logic". The Soviet reluctance to explore space beyond Earth orbit has already been mentioned. That shouldn't affect what America chose to do—but of course it did. What's the point of a race to Mars if the other side doesn't want to compete? In October 1969, just three months after Apollo 11, the Soviet leader Leonid Brezhnev made it clear that his country's space interests lay solely in Earth-orbiting space stations:

> Our way to the conquest of space is the way of solving vital, fundamental tasks—basic problems of science and technology. Our science has approached the creation of long-term orbital stations and laboratories as the decisive means to an extensive conquest of space. Soviet science regards the creation of orbital stations with changeable crews as the main road for man into space. [48]

The Soviets were true to their word. Soyuz, which might at one time have taken cosmonauts to the Moon, now had a different role: to ferry them up to orbiting space stations. The first such stations—called Salyut, and very modest in scale—appeared in the 1970s and early 80s. They were followed by the larger Mir, which was launched in 1986 and outlasted the Soviet Union itself, remaining in use until 1999.

Against this background, NASA had very little incentive to go to Mars—or even to strengthen its presence on the Moon. A cynic might say they'd only gone there in the first place to prove they were better than the Russians, or as Frank Borman put it: "the Apollo programme was just a battle in the Cold War" [40]. That battle had now degenerated—on both sides—into a tactical retreat into low Earth orbit. While the Soviets played with Salyut and Mir, NASA pressed ahead, in rather desultory fashion, with the Space Shuttle.

Politicians, however, weren't the only ones to blame for the winding-down of the space programme. Public opinion, too, was turning against it. After Apollo 11, there was a rapid decline in interest in the later Moon landings—of which there were six between 1969 and 1972, although not everyone noticed at the time. America's Vice-President during these years, Spiro Agnew, was one of the few politicians with a genuine enthusiasm for space exploration. It wasn't a view that made him popular, as this quote from Stephen Baxter's afterword to *Voyage* makes clear:

Agnew himself was … a champion within the White House of aiming for Mars—even though he was booed when he spoke of the project in public. [49]

It's ironic that the public's growing disillusionment with real-world space travel coincided with a huge upsurge in the popularity of space-based science fiction. For most of the 1950s and 1960s, the genre had been limited to magazines and paperback books—and interest was confined to a relatively small community of fans. The star attractions at SF conventions—which boasted three-digit attendance figures, not the six-digit ones of today's media-driven extravaganzas—were writers like Asimov and Heinlein, not A-list Hollywood celebrities. Even the original version of *Star Trek*, which first appeared on TV screens in the late 60s, failed to achieve decent viewing figures—and it was cancelled after just three seasons. The very last episode aired on 3 June 1969, just six weeks before Apollo 11 landed on the Moon.

A decade later, it was the turn of real-world space missions to be cancelled due to lack of interest, while audiences queued round the block to see the latest space adventure movies. The big change came with the release of *Star Wars* in 1977, followed by the first big-screen incarnation of *Star Trek* two years later. Other franchises followed—and by the end of the Cold War in 1991, "blockbuster movie" was almost synonymous with "sci-fi".

One person who spotted the trend almost immediately was General Sir John Hackett, whose novel *The Third World War* was published in 1979. Told from the future perspective of an East-West war in 1985, it describes the situation in the late 70s as follows:

As Congress cut back the funds for NASA's space programme the Earthbound mortals with their daily lives to lead rather lost interest. Films like the record running *Star Wars*, first screened in 1977 and still showing in London at the outbreak of the war, and books about interstellar wars in deep space, fascinated and absorbed the public while real men and their machines performing tasks as they orbited the world, no further away than London is from Manchester in England, seemed of no particular account. [50]

As the Cold War wound down towards its close, NASA's flagship technology was not the Mars exploration vehicle that might have been, but the Space Shuttle. This was essentially a large aircraft that could be launched into space like a rocket, orbit the Earth for a week or two, and then re-enter the atmosphere and glide to a landing. It was only in this last phase that the aircraft-like design was really necessary—but it was critical to the Shuttle's success. For that reason, a preliminary test vehicle was built that was airworthy but not spaceworthy—its role being to make sure the Shuttle performed the way it was supposed to during the gliding phase.

When the test vehicle was unveiled in September 1976, the name painted on its side was "Enterprise"—just like the famous starship in *Star Trek*. The similarly of names wasn't a coincidence, as the NASA website explains:

> Enterprise, the first Space Shuttle Orbiter, was originally to be named Consti-tution (in honour of the US Constitution's bicentennial). However, viewers of the popular TV science fiction show *Star Trek* started a write-in campaign urging the White House to select the name Enterprise. [51]

The actual naming decision was made at quite a late stage in the proceedings by President Gerald Ford—as previously classified memoranda released in 2014 reveal. In one of them, an advisor tells Ford:

> NASA has received hundreds of thousands of letters from the space-oriented *Star Trek* group asking that the name "Enterprise" be given to the craft. This group comprises millions of individuals who are deeply interested in our space programme. . .. Use of the name would provide a substantial human-interest appeal to the rollout ceremonies scheduled for this month in California. [52]

Just what that "human-interest appeal" amounted to can be seen in Fig. 11, which shows *Star Trek*'s creator Gene Roddenberry together with several cast members at the shuttle rollout ceremony. It was a triumph, of a kind, for the science fiction community—although most SF fans would probably have preferred to see NASA going to Mars instead.

Fig. 11 *Star Trek* creator Gene Roddenberry (near the centre, in a brown suit) and some of the cast members at the 1977 unveiling of the first test version of the Space Shuttle—which was called Enterprise after the fictional starship in that TV series (NASA image)

References

1. S. D. Tucker, *Space Oddities* (Amberley, Stroud, 2017), pp. 134, 135
2. B. Rickard, The First Forteans, Part 2, in *Fortean Times* (Christmas 2013), pp. 50, 51
3. W. Ley, The Dawn of the Conquest of Space, in *Astounding Stories* (March 1937), pp. 104–110
4. N. Watson, Reaching for the Stars, in *History Today* (January 2013), http://www.historytoday.com/nigel-watson/reaching-stars
5. J.L. Ingham, *Into Your Tent* (Plantech, Reading, 2009), p. 78
6. C. Bennett, Rocket in his Pocket, in *Fortean Times* (March 2000), pp. 34–38
7. N. Redfern, *Science Fiction Secrets* (Anomalist Books, San Antonio, 2009), p. 100
8. A.C. Clarke, *Astounding Days* (Gollancz, London, 1990), p. 191
9. A.C. Clarke, *Astounding Days* (Gollancz, London, 1990), p. 152
10. A.C. Clarke, *Profiles of the Future* (Pan Books, London, 1973), p. 26
11. A.C. Clarke, *1984: Spring* (Granada, London, 1984), pp. 146, 147

12. T. Burnett, *Who Really Won the Space Race?* (Collins & Brown, London, 2005), p. 30

13. T. Burnett, *Who Really Won the Space Race?* (Collins & Brown, London, 2005), pp. 17–19

14. G. Harry Stine, *ICBM: The Making of the Weapon that Changed the World* (Orion Books, New York, 1991), p. 61

15. H.S. Truman, Statement by the President Announcing the Use of the A-Bomb at Hiroshima (6 August 1945), *Harry S. Truman Library and Museum*, https://www.trumanlibrary.org/publicpapers/index.php?pid=100

16. R. Jackson, *Encyclopedia of Military Aircraft* (Paragon, Bath, 2002), p. 260

17. J.W. Campbell., Not Quite Rockets, in *Astounding Science Fiction* (April 1944), pp. 99–103

18. G. Harry Stine, *ICBM: The Making of the Weapon that Changed the World* (Orion Books, New York, 1991), p. 105

19. H. Gernsback, Stations in Space, in *Air Wonder Stories* (April 1930), p. 869

20. A.C. Clarke, *Voices from the Sky* (Mayflower, London, 1969), p. 105

21. R.A. Heinlein, *The Man Who Sold the Moon* (SF Gateway, Kindle edition, 2014), loc. 1250

22. R.A. Heinlein, *Destination Moon* (Heinlein Archives, Kindle edition, 2013), loc. 35

23. R.A. Heinlein, *Destination Moon* (Heinlein Archives, Kindle edition, 2013), loc. 671

24. I. Fleming, *Dr. No* (Pan Books, London, 1960), pp. 142–143

25. P. Drye, *False Steps: The Space Race as it Might Have Been* (Baggage Books, Kindle edition, 2015), loc. 138

26. W. Tenn, *Project Hush* (Project Gutenberg, 2010), http://www.gutenberg.org/files/32654/32654-h/32654-h.htm

27. T. Burnett, *Who Really Won the Space Race?* (Collins & Brown, London, 2005), pp. 205–206

28. T. Burnett, *Who Really Won the Space Race?* (Collins & Brown, London, 2005), p. 236

29. G. Harry Stine, Sputnik: One Reason Why We Lost, in *Magazine of Fantasy and Science Fiction* (January 1958), p. 75

30. Lester del Rey, Behind the Sputniks, in *Fantastic Universe* (April 1958), pp. 56–65

31. T. Burnett, *Who Really Won the Space Race?* (Collins & Brown, London, 2005), p. 7

32. J.F. Kennedy, *Special Message to the Congress on Urgent National Needs* (25 May 1961), https://www.nasa.gov/vision/space/features/jfk_speech_text.html

33. R.A. Heinlein, *Rocketship Galileo* (New English Library, London, 1971), p. 31

34. U.S. Army, *Project Horizon, Volume I: Summary and Supporting Considerations*, http://nsarchive.gwu.edu/NSAEBB/NSAEBB479/docs/EBB-Moon01_sm.pdf

35. D. Kyle, *A Pictorial History of Science Fiction* (Hamlyn, London, 1976), p. 88

36. Paul Drye, *False Steps: The Space Race as it Might Have Been* (Baggage Books, Kindle edition, 2015), loc. 834
37. Air Force Space Systems Division, *Lunar Expedition Plan* (1961), http://www.astronautix.com/data/lunex.pdf
38. A.C. Clarke, *Astounding Days* (Gollancz, London, 1990), p. 148
39. P. Drye, *False Steps: The Space Race as it Might Have Been* (Baggage Books, Kindle edition, 2015), loc. 1528
40. M. Klesius, To Boldly Go, in *Smithsonian Air and Space Magazine* (December 2008), https://www.airspacemag.com/space/to-boldly-go-133005480/
41. J. Sutton, *First on the Moon* (Project Gutenberg, 2013), http://www.gutenberg.org/ebooks/43235
42. J. Sutton, *Apollo at Go* (Mayflower, London, 1964), p. 20
43. J. Sutton, *Apollo at Go* (Mayflower, London, 1964), p. 88
44. NASA, *Apollo Lunar Surface Journal*, Apollo 11 landing, https://www.hq.nasa.gov/alsj/a11/a11.landing.html
45. J. Sutton, *Apollo at Go* (Mayflower, London, 1964), p. 91
46. S. Knoll, First Man on the Moon Has TV Networks in Orbit, in *Variety* (July 1969), http://variety.com/1969/biz/news/first-man-on-the-moon-has-tv-networks-in-orbit-1201342630/
47. Stephen Baxter, *Voyage* (Harper Voyager, London, 1996), p. 585
48. B. Evans, The Main Road: The World's First Space Station, in *America Space* (April 2013), http://www.americaspace.com/2013/04/20/the-main-road-the-worlds-first-space-station-part-1/
49. Stephen Baxter, *Voyage* (Harper Voyager, London, 1996), p. 588
50. G.S.J. Hackett, *The Third World War* (Sphere, London, 1979), p. 253
51. Kennedy Space Centre, Enterprise(OV-101), https://science.ksc.nasa.gov/shuttle/resources/orbiters/enterprise.html
52. M. Strauss, Declassified Memos Reveal Debate Over Naming the Shuttle Enterprise, io9 (July 2014), https://io9.gizmodo.com/declassified-memos-reveal-debate-over-naming-the-shuttl-1603073259

Electronic Brains

In which science fiction, having successfully foreseen atomic power and space travel, completely misses the third great technological revolution of the 20th century: the advent of miniaturized solid-state electronics. Instead, its predictions of sentient humanoid robots and giant, building-sized computers—often bent on world conquest—failed to materialize (in the Cold War timeframe, anyway). In more general terms, however, the science-fictional vision of automated warfare was borne out in the guided missiles and defence computers of the Vietnam era.

Robot Dreams

Many of the words most commonly associated with science fiction were actually coined earlier, outside the genre. The word "android", for example, was used to refer to human-shaped automata as long ago as the 18th century. The use of "telepathy" to mean direct mind-to-mind communication dates back to the psychic researchers of the 19th century, while the word "teleportation" was used by Charles Fort in the 1930s to refer to objects mysteriously disappearing from one location and reappearing at another. As for that archetypally sci-fi word "cybernetics"—it was coined by the mathematician Norbert Wiener in 1948 to describe the scientific study of control systems.

The word "robot", on the other hand, really does come from fiction—as scientist and SF fan Stephen Webb relates:

© Springer International Publishing AG, part of Springer Nature 2018   **81**
A. May, *Rockets and Ray Guns: The Sci-Fi Science of the Cold War*, Science and Fiction,
https://doi.org/10.1007/978-3-319-89830-8_3

The first use of the word "robot" came in the play *R.U.R.* (1920), by the Czech writer Karel Čapek. The acronym in the title stands for "Rossum's Universal Robots". The action starts in a factory that makes robots—artificial people—to do the hard, manual work that humans naturally enough dislike. Indeed, Čapek derived the word "robot" from the Czech word *robota*, which means statute labour; he was making the point that the robots were slaves. In the play the robots eventually rebel, and humans become extinct. [1]

Within a few years, "robot" had become part of the language, and not just in the literal sense of an artificial humanoid. It was used figuratively too. An early example can be found in Agatha Christie's detective novel *Three Act Tragedy* (1935), in which one of characters—explaining how self-controlled his secretary is—says "Miss Milray's the perfect robot" [2].

By this time, robots had made the transition from the highbrow literature of Čapek's play to the lowbrow SF of the pulp magazines—and from organically synthesized humans to clanking, metallic machines. One of the first pulp stories to feature this more familiar concept of a robot, in the October 1929 issue of *Air Wonder Stories*, was "The Robot Master" by O. Beckwith (see Fig. 1). A typical "mad scientist" story, it features an army of remote-controlled robots created with the aim of conquering the world:

> "Look here," said the professor. He was leading the way to a massive board on the right side of the room. "Here is the robot control board. Do you see these many hundreds of tiny, mirror-like circles? On those register all the things that pass before the eye in my robots. . . . And when I control all of them at the same time . . . can you doubt now that I can conquer the world?" [3]

A word derived from robot is "robotics", referring to the branch of engineering concerned with the construction of autonomous mechanical systems. It's a perfectly respectable subject; the Robotics and Automation Society was founded by the prestigious Institute of Electrical and Electronics Engineers in 1987. The word, however, was coined much earlier than that, in a distinctly less academic context. The word "robotics" made its first appearance in Isaac Asimov's short story "Liar!", in the May 1941 issue of *Astounding Science Fiction*.

It was in a slightly later story—"Runaround", in the March 1942 *Astounding*—that Asimov first articulated his three laws of robotics. The second and third of these are uncontroversial, embodying the idea that a robot should obey instructions, and avoid any action likely to result in damage to itself—both

Fig. 1 Illustration for "The Robot Master" in *Air Wonder Stories*, October 1929—one of the first pulp magazine stories to adopt Karel Čapek's word "robot" (public domain image)

obvious prerequisites for any autonomous machine. The real interest, however, lies in Asimov's first law, which overrides the other two:

> A robot may not injure a human being, or, through inaction, allow a human being to come to harm. [4]

In some situations, this makes sense. A self-driving car, for example, ought to stop at a pedestrian crossing if there are people on it, even if the light is green. But what about a military robot—or "drone", in present-day terminology? If it had been programmed to bomb an enemy command post, it would be ludicrous to think it might disobey that order if there were people inside. Yet that is what Asimov's first law implies.

One of the first people to consider the ethics of military robots was the famous 20th century philosopher Bertrand Russell—and he chose to do so in the form of a science fiction story. "Dr Southport Vulpes's Nightmare: The Victory of Mind Over Matter" appears in his short story collection *Nightmares of Eminent Persons*, published in 1954 when Russell was over 80. The eponymous Dr Southport Vulpes—a scientist at the Ministry of Mechanical Production—dreams about a future war that is fought solely by his mechanical creations:

> The behaviour of our robots is in all respects better than that of the accidental biological product which has hitherto puffed itself up with foolish pride. How ingenious are their devices! How masterly their strategy! How bold their tactics, and how intrepid their conduct in battle! . . .
>
> The rival armies met in their millions. The rival planes, guided by robots, darkened the sky. Never once did a robot fail in its duty. Never once did it flee from the field of battle. Never once did its machinery whirr in response to enemy propaganda. [5]

The idea of a mechanical army substituting for a human one is daunting enough. A year before Russell's story, however, Philip K. Dick came up with an even more sinister use for military robots. In "Second Variety" (1953), he envisioned killer robots so life-like that they could mingle undetected among humans. The first such robot encountered by the protagonist—an American soldier named Hendricks—is modelled after a harmless-seeming boy who calls himself David. Set, like so much of Dick's fiction, in a future extrapolation of the East-West Cold War, Hendricks' life is saved when a Russian soldier shoots the robot (see Fig. 2):

> "Why did you do it?" he murmured thickly. "The boy."
>
> "Why?" One of the soldiers helped him roughly to his feet. He turned Hendricks around. "Look." . . .
>
> Hendricks looked. And gasped.
>
> "See now? Now do you understand?"
>
> From the remains of David a metal wheel rolled. Relays, glinting metal. Parts, wiring. One of the Russians kicked at the heap of remains. Parts popped out, rolling away, wheels and springs and rods. A plastic section fell in, half charred. Hendricks bent shakily down. The front of the head had come off. He could make out the intricate brain, wires and relays, tiny tubes and switches, thousands of minute studs—
>
> "A robot," the soldier holding his arm said. [6]

Fig. 2 Illustration for Philip K. Dick's short story "Second Variety" in *Space Science Fiction*, May 1953 (public domain image)

For modern readers, Dick's reference to "relays" and "tubes" may need some explanation. "Relay", in this context, refers to an electrical switch that can be opened or closed by applying current to an electromagnet. Large numbers of such relays were used in early computers to perform mathematical operations. Another way of doing the same thing is by controlling, not the clicking of a mechanical switch, but the flow of electrons in an electronic circuit. In the first half of the 20th century, the only device that could do this was the vacuum tube, or "thermionic valve" in British English. Like a relay, such a tube can be used as an electrically operated switch, or alternatively as an amplifier or oscillator.

Tubes and relays were physically large objects—with dimensions measured in centimetres—so it would be impossible to fit more than a few hundred of them into a space the size of a human skull. For that reason, Dick's suggestion that a robot built from such components might be able to talk and act just like a real human is pure fiction.

Aware of this problem, Isaac Asimov avoided using the word "electronic" in his own robot stories. He adopted a slightly different word instead: "positronic". As exciting as that sounds, it's scientifically meaningless in this context. The positron is a subatomic particle, first discovered in 1932, that is almost identical to an electron except that it has positive rather than negative charge. It's an example of antimatter—extremely difficult to create in the laboratory, and with the habit of disappearing in a flash of radiation as soon as it comes into contact with ordinary matter. For Asimov, however, "positron" was simply a scientific-sounding term that was highly fashionable at the time—rather like "Higgs boson" today. As science writer Brian Clegg explains:

> Positrons were very much in the science news in the 1940s . . . so they turned up in the positronic brains that were a feature of Isaac Asimov's robots. [7]

Several decades after they were written, Asimov reprinted some his positronic robot stories in a collection entitled *Robot Dreams* in 1986. The same book also contains several stories of a similar vintage featuring a computer called Multivac. Unlike Asimov's robots, the latter was merely electronic, not positronic. Conceived as it was in the 1950s, that meant Multivac—filled with clicking relays and glowing vacuum tubes—was necessarily huge. Here is an excerpt from one of the most famous Multivac stories, "The Last Question" (1956)—depicting the supposed state-of-the-art in 2061:

> Alexander Adell and Bertram Lupov were two of the faithful attendants of Multivac. As well as any human beings could, they knew what lay behind the cold, clicking, flashing face—miles and miles of face—of that giant computer. They had at least a vague notion of the general plan of relays and circuits that had long since grown past the point where any single human could possibly have a firm grasp of the whole. [8]

By the time *Robot Dreams* appeared in 1986, everyone knew that 21st century computers wouldn't be cumbersome, clunky behemoths like Asimov's Multivac. After their tremendous success with atom bombs and space rockets—and their over-ambitious prophecies about humanoid robots—SF writers completely failed to anticipate the future trend of computer evolution.

Asimov's depiction of Multivac was based on a future extrapolation of the very first electronic computers. Built from relays and vacuum tubes, these had been used almost exclusively for mathematical computations (hence the name). One of the first digital computers—the Electronic Numerical Integrator and Computer, or ENIAC—had been developed by the US Army's Ballistic Research Laboratory in 1946. Among other things, ENIAC was used to help optimize the design of nuclear warheads. It also shaped the way that Isaac Asimov and other authors thought about computers. As Stephen Webb explains:

> The SF authors who marvelled in 1946 at ENIAC, the first electronic general-purpose computer, were taken in by the sheer size of this "giant brain": the thousands of vacuum tubes and diodes, relays and resistors, filled a large room. [9]

Ironically, however, this was one point in history where the real world was way ahead of science fiction. Even as Asimov was writing his Multivac stories, a revolution was in progress that would render them obsolete within a few years.

The Silicon Revolution

Science fiction has a long-established fascination with the chemical element silicon—but not for the reason it's so important today. As John W. Campbell explained in the February 1945 issue of *Astounding Science Fiction*, in an editorial entitled "Silicon for Carbon":

> All life, as we know it, is based on the immensely complex chemistry of carbon, hydrogen, oxygen and nitrogen compounds, plus minor additions of other elements—but carbon is the great *sine qua non*. The quadrivalent carbon atom, with its ability to link in chains, makes possible those complex molecules. Only two other elements remotely approach carbon's ability to form long, linked chains—boron and silicon. Both have been proposed, rather haltingly, as possible substitutes for carbon in the life processes of an evolution of animals on extraterrestrial planets in various science fiction stories. [10]

Of the two carbon-like elements mentioned by Campbell, boron is very rare, both on Earth and in the wider universe—but silicon is one of the commonest there is. Here on Earth, it's even more abundant than carbon. In the form of silica, or silicon dioxide, it's present in most rocks. One of the first stories to suggest that silicon, rather than carbon, might form the basis of an extraterrestrial life-form was Stanley G. Weinbaum's "A Martian Odyssey" (1934):

The beast was made of silica! There must have been pure silicon in the sand, and it lived on that. Get it? We . . . are carbon life; this thing lived by a different set of chemical reactions. It was silicon life! [11]

The idea of rock-like silicon-based creatures took hold. Later and better known examples include the Silicony in Isaac Asimov's "The Talking Stone" (1955) and the Horta in the first-season *Star Trek* episode "The Devil in the Dark" (1967).

All this was a red herring, though. In today's world, silicon is one of the most important elements there is—but not because of its carbon-like chemical structure. Instead, the unique significance of silicon comes from something the SF writers overlooked—its strange electrical properties.

Most materials either conduct electricity very easily, like copper wires, or very poorly, like the ceramics and plastics used as electrical insulators. Silicon, on the other hand, is a "semiconductor". It will pass an electric current, but not as freely as a true conductor like copper. In itself, that's not very exciting—but silicon's conductivity can be adjusted by adding impurities to it. That's a property that's exploited in one of the most important inventions of the 20th century: the transistor.

A transistor functions like a miniaturized, solid-state version of the clunky old vacuum tube. In just the same way, it can be used to control the flow of current in an electronic circuit—but on a physical scale that is almost unimaginably smaller. Unlike vacuum tubes, which are discrete components, a large number of transistors can be etched onto a single chip of silicon. At the time of writing (2018), a state-of-the-art microprocessor may contain as many as a billion transistors—and that's a number that will undoubtedly rise in the future.

Given its lack of prominence in the science fiction of the 1940s and 1950s, it's surprising to learn that the transistor was invented—by John Bardeen, Walter Brattain and William Shockley of Bell Laboratories—as long ago as 1947. One of the few members of the SF community who took notice of it at the time was the engineer John R. Pierce—who, as mentioned in the previous chapter, also wrote SF under the pen-name of J. J. Coupling. That was the byline he used for a non-fiction article in the June 1952 *Astounding*, in which he berated his fellow writers for ignoring the transistor:

Authors don't invent minuscule electronic gadgets which oscillate, amplify and perform all the functions of a vacuum tube and yet take almost no space and even less power. And don't try to make such things mysterious by having them featureless blobs of crystal. . . . Readers, if that sort of thing interests you, turn from the unimaginative fictioneers right straight around to the technology of today, and hear about transistors. Transistors were announced publicly about four years ago by the Bell Telephone Laboratories. They were invented by J. Bardeen and W. H. Brattain. The research which led to their invention was part of an extensive programme carried out under the direction of W. Shockley. [12]

Pierce then goes on to ascribe credit for the very word "transistor" to himself:

The name was suggested by two other solid-state devices, the varistor and the thermistor, and was supplied by an obscure character, J. R. Pierce. [12]

As surprising as this assertion is, even Wikipedia appears to go along with it [13]. In a roundabout way, therefore, the SF community can claim a starring role in the silicon revolution after all. In any case, just as Pierce implied earlier, transistors are every bit as weird and wonderful as any fictitious "magic crystal". Inexplicable in terms of classical physics, it's necessary to call on the strange surrealism of quantum theory to explain them properly. In the words of Brian Clegg:

The operation of semiconductors is a purely quantum effect that, unlike ordinary electricity, could not even be understood without a grasp of quantum theory. Valves were quantum devices, but they could be built and operated without realizing what was going on. . ... With the introduction of transistors, we saw the first technology emerging that required an understanding of quantum physics to design it. [14]

Transistorized electronics—the principal technology of the modern world—emerged just a few years after the first nuclear weapons—the principal technology of the Cold War. The two, however, have always had an uneasy relationship. Of course, nuclear missiles are dependent on electronic systems for targeting and control, but those same electronic systems really don't like nuclear explosions.

One of the less well-known effects of such explosions—on top of the blast and radiation—is something called an electromagnetic pulse, or EMP. It turns out transistors are particularly vulnerable to this—much more so than a vacuum tube or valve. The situation was neatly explained, at the height of the Cold War in 1984, by Australian science popularizer Karl Kruszelnicki:

EMP was first noticed by US scientists back in 1962, when they had a little nuclear experiment called Starfish Prime. They exploded a 1.5 MT nuclear weapon 400 km above the ground level of Johnston Island in the Pacific. 1500 km away in Hawaii there was massive electronic destruction as 300 street lights blew up, burglar alarms triggered off, power lines fused and TV sets exploded. The EMP will burn out transistors and integrated circuits, it will burn out radios and it will burn out TVs. Radio valves are a billion times more resistant to EMP than integrated circuits. The Russians know about this, and in their MiG-25 Foxbat interceptor fighter they use—not only valves to stop EMP—but also a double skin on the aircraft to stop electromagnetic radiation penetrating. United States investigators found this in 1976 when a Soviet pilot defected to Japan and they pulled the plane to pieces. They started laughing because they thought "valves in 1976, how primitive!" But late in 1977 the Pentagon said "Yes, go ahead and use valves, they are a better device." [15]

What this meant was that, as the world became increasingly reliant on silicon-based electronics, the idea of fighting a nuclear war became less and less credible. The general public might have been comforted by this fact, but by and large they were unaware of it. As for the military—it threatened to ruin all their carefully made plans. For them, the ideal solution would be an electronic chip that was resistant to EMP.

Just such a device—a purely fictional one—features in the 1985 James Bond movie *A View to a Kill*. The following exchange takes place between Bond and his science advisor, known only by the code-name "Q":

Q: Until recently, all microchips were susceptible to damage from the intense magnetic pulse of a nuclear explosion. . ...

James Bond: Yes . . . one burst in outer space over the UK and everything with a microchip in it, from, well, the modern toaster to the most sophisticated computers in our defence systems would be rendered absolutely useless. . ...

Q: One of our private defence contractors came up with this: a chip totally impervious to magnetic pulse damage. [16]

Giant Computers

Was it because gigantic, vacuum-tube-filled computers would be less vulnerable to EMP that SF writers continued to write about them long after the invention of the transistor? Probably not. In hindsight, however, it provides a charitable way to view these stories—set as they so often are against the backdrop of the Cold War, with its ever-present threat of nuclear war.

While the robot workers envisaged by Karel Čapek were given menial tasks to perform, the giant electronic brains of later fiction occupied the other end of the value chain, often being entrusted with top level administrative work. The consequences for humanity, however, tended to be every bit as disastrous.

A good example is Philip K. Dick's 1960 novel, *Vulcan's Hammer*. Near the start, a history teacher tells to her class how, following the conclusion of "Atomic War One" in 1992:

> The world-wide Unity organization formally agreed that the great computer machines developed by Britain and the Soviet Union and the United States, and hitherto used in a purely advisory capacity, would now be given absolute power over the national governments in the determination of top-level policy. [17]

Not surprisingly—given that this is a work of fiction—one of the computers, Vulcan 3, decides to seize the opportunity and take over the world. As matters start to get out of hand, a government official ventures inside the building-sized computer:

> Very little of the computer was visible; its bulk disappeared into regions which he had never seen, which in fact no human had ever seen. During the course of its existence it had expanded certain portions of itself. To do so it had cleared away the granite and shale earth; it had, for a long time now, been conducting excavation operations in the vicinity. . . . Beneath his feet the floor vibrated. This was the deepest level which the engineers had constructed, and yet something was constantly going on below. He had felt vibrations before. What lay down there? No black earth; not the inert ground. Energy, tubes and pipes, wiring, transformers, self-contained machinery. He had a mental picture of relentless activity going on: carts carrying supplies in, wastes out; lights blinking on and off; relays closing; switches cooling and reheating; worn-out parts replaced; new parts invented; superior designs replacing obsolete designs. And how far had it spread? Miles? . . . Did it go down, down, forever? [18]

At the time the novel was written, there was some justification for the fears it represented. In the late 1950s, the North American Air Defence Command (NORAD) had delegated certain aspects of military decision-making to a computer network called SAGE: the Semi-Automatic Ground Environment. A fictionalized, first-person account of this, called "I, SAGE", appeared in comic-book form in the first issue of *Atom-Age Combat*, cover-dated February 1958 (see Fig. 3).

As real-world history unfolded, SAGE didn't go mad and try to enslave humanity. That did, however, happen in the 1970 film *Colossus: The Forbin*

Fig. 3 Two panels from the story "I, SAGE" in the first issue of *Atom-Age Combat*, from February 1958—a reasonably factual account of "the electronic system on which this country depends to deal with the threat of enemy bombers" (public domain image)

Project—whose eponymous computer, Colossus, is clearly modelled on SAGE. Describing the movie as "the granddaddy of all computer run amok films", the back cover of the 2017 DVD provides the following summary:

> When computer genius Charles Forbin creates a massive computer complex that is capable of independently regulating the national defence of the United States, it appears that no enemy will ever be able to penetrate its sovereign borders. But ... it is discovered the Russians have built an equally sophisticated computer and that these two machines have linked, sharing classified information and national secrets. Desperately Forbin and his Soviet counterparts try to stop the all-knowing computers from seizing command of the world's nuclear missile stockpiles. [19]

That's just the start of it. By the end of the film, Colossus has become the electronic equivalent of George Orwell's Big Brother:

> This is the voice of world control. I bring you peace. It may be the peace of plenty and content or the peace of unburied death. The choice is yours: obey me and live, or disobey and die. . ... We can coexist, but only on my terms. You will say you lose your freedom. Freedom is an illusion. [20]

The real world SAGE was, of course, much more limited than its cinematic counterpart. It merely combined data from multiple sources in order to simplify matters for human decision makers—it didn't make any decisions of its own. Nevertheless, the idea that it might have done wasn't as science-fictional as it sounds—even in the 1950s when SAGE was built. It had been known since World War Two that many militarily-important decisions are amenable to mathematical analysis, and methods for handling them had been developed under the vague and innocuous-sounding name of "operational research".

Arthur C. Clarke drew on this subject for his 1956 short story "The Pacifist", later collected in *Tales from the White Hart*. Like most of the stories in that book, it's narrated by a character named Harry Purvis. This particular tale combines the real-world discipline of operational research with yet another fictional super-computer:

"As you all know," began Harry, "Science with a capital S is a big thing in the military world these days. The weapons side—rockets, atom bombs and so on—is only part of it, though that's all the public knows about. Much more fascinating, in my opinion, is the operational research angle. You might say that's concerned with brains rather than brute force. I once heard it defined as how to win wars without actually fighting, and that's not a bad description. Now you all know about the big electronic computers that cropped up like mushrooms in the 1950s. Most of them were built to deal with mathematical problems, but when you think about it you'll realize that war itself is a mathematical problem. It's such a complicated one that human brains can't handle it—there are far too many variables. . . .. But a machine—that would be a different matter." [21]

In the story, the giant computer built to deal with such problems is called Karl, after the great military strategist Karl von Clausewitz. Unfortunately, the task of programming Karl is given to a pacifist mathematician, Dr Milquetoast (that's not his real name, but it's what Purvis calls him). Milquetoast makes sure the computer is constitutionally incapable of answering any question put to it by the General in charge of the project:

Somewhere tucked away in Karl's capacious memory units was a superb collection of insults, lovingly assembled by Dr Milquetoast. He had punched on tape, or recorded in patterns of electrical impulses, everything he would like to have said to the General himself. But that was not all he had done; that would have been too easy, not worthy of his genius. He had also installed what could only be called a censor circuit—he had given Karl the power of discrimination. Before solving it, Karl examined every problem fed into him. If it was concerned with pure mathematics, he cooperated and dealt with it properly. But if it was a military problem—out came one of the insults. [21]

Cuban Missile Crisis Game Tree
A: The United States (President Kennedy)
B: The Soviet Union (Premier Khrushchev)

Fig. 4 Game theory applied to the Cuban Missile Crisis (Wikimedia user John Yaeger, CC-BY-SA-4.0)

Although the story is fiction—and humorous fiction at that—there's nothing fundamentally wrong with the idea of using a computer to solve problems of military strategy. One way to do this is through a branch of operations research with the unlikely-sounding name of "game theory". In colloquial usage, a game is something that isn't meant to be taken seriously—but that's not the way mathematicians use the word. For them, a game is simply a confrontation between two opposing players, each with a clearly defined objective. The aim of game theory is to calculate each player's optimum strategy—and the result applies just as well to a nuclear confrontation as to a game of poker.

That's illustrated in Fig. 4, which shows a simplified "game tree" for the Cuban missile crisis of October 1962. One of the tensest moments of the Cold War, this occurred when Khrushchev, the Soviet leader, deployed a number of nuclear missiles to the island of Cuba, within easy reach of the United States. The Americans, under President Kennedy, were determined to see the missiles removed, while Khrushchev was equally determined to keep them there. The diagram shows the basic options open to each side.

Given the various options, and the strategic objectives of each side, the equations of game theory work out which is the best option to pursue. A tacit assumption is made that both players are equally intelligent and equally rational. That means, for example, that player A can count on player B acting in a way that provides maximum benefit to B. So A's best option is whichever strategy *minimizes* that maximum benefit to B. This convoluted logic is called the "minimax" principle—and it's referred to by Philip K. Dick in the preface to his novel *Solar Lottery* (1955):

> I became interested in the theory of games, first in an intellectual manner (like chess) and then with a growing uneasy conviction that Minimax was playing an expanding role in our national life. . ... Both the US and the Soviet Union employ Minimax strategy as I sit here. While I was writing *Solar Lottery*, von Neumann, the co-inventor of the games theory, was named to the Atomic Energy Commission, bearing out my belief that Minimax is gaining on us all the time. [22]

As Dick says, one of the key people in the development of game theory was John von Neumann. That's a name that cropped up a couple of times in the first chapter, "The Super-Bomb"—first as a member of the Manhattan Project, and then as the originator of the term "mutual assured destruction". It was also on von Neumann's initiative that the ENIAC computer, mentioned earlier in this chapter, was used in the design of some of the earliest nuclear weapons. Isaac Asimov, after referring to von Neumann's nuclear work, continues as follows:

> Even more important was his development of a new branch of mathematics called "game theory". . ... This branch of mathematics is called game theory because it works out the best strategies to follow in simple games, such as coin matching. However, the principles will apply to far more complicated games such as business and war, where an attempt is made to work out the best strategy to beat a competitor or an enemy. . ... Von Neumann also applied his mathematical abilities to directing the construction of giant computers, which in turn performed high-speed calculations that helped in the production of the H-bomb and in reducing it to a size small enough to be fired by missile (some visualize a future in which war is fought not only by the pressing of buttons, but by means of a computer working out the equations of game theory and itself pushing the buttons). [23]

That last parenthetical suggestion may sound familiar to fans of *Star Trek*, since it's the way they wage war on the planet Eminiar in the first season episode "A Taste of Armageddon" (1967). Here's a scene where Captain Kirk and Mr Spock confront a local dignitary, called Anan:

> Kirk: I've been in contact with my ship, which has had this entire planet under surveillance. All during this so-called attack of yours, we have been monitoring you. There's been no attack, no explosions, no radiations, no disturbances whatsoever. If this is some sort of game you're playing . . .

Anan: This is no game, Captain. Half a million people have just been killed. . ... Launch immediate counter attack.

Spock: Computers, Captain. They fight their war with computers. Totally.

Anan: Yes, of course.

Kirk: Computers don't kill a half million people.

Anan: Deaths have been registered. Of course they have 24 hours to report.

Kirk: To report?

Anan: To our disintegration machines. [24]

Drone Wars

The concept of a "drone"—an autonomous vehicle for military or civilian use—is often seen as a modern one, post-dating the Cold War. That's because the micro-miniaturized electronics that make present-day drones so effective is a relatively recent development. The basic idea, however, goes back much further than that—as science journalist David Hambling explains:

> In some ways the true inventor of unmanned warfare was Nikola Tesla, who demonstrated a miniature boat controlled by radio waves at Madison Square Garden in 1898. Tesla believed that a version armed with torpedoes could sink battleships and lead to a new age in which wars were fought between machines with no human combatants. As with many of his projects, Tesla never developed the idea beyond the initial demonstration. [25]

Tesla was a prolific inventor who we will meet several more times before the end of this book. In many instances—drones among them—his inventions were decades ahead of their time. Although Tesla had demonstrated their basic feasibility before the start of the 20th century, it was the middle of that century before anything remotely like a drone entered military service. When it did, it was called a guided missile.

The first experimental guided missiles appeared in World War Two, but it was the 1950s before they began to enter operational service in any numbers. Unlike a ballistic missile, which is launched on a pre-calculated trajectory towards a fixed target, a guided missile is designed to attack a moving target. That means it has to do its own navigation, which is where the "drone-like" behaviour comes in. By the time of the Vietnam war in the 1960s, numerous types of guided missile were in service, characterized by the way they were launched and the intended target—air-to-air, air-to-surface or surface-to-air, for example.

To know where to go, a guided missile needs on-board sensors. Tradition-ally, science fiction depicted robots as having similar senses to humans—making use of sound or visual optics, for example. Real world guided missiles tend to use different "senses". One well-known option is the heat-seeking missile, which homes in on the infrared radiation generated by the target itself. This was the principle employed, during the Vietnam war, by the US Side-winder air-to-air missile—and its North Vietnamese equivalent, the Soviet-built K-13, which was so similar to the Sidewinder it may have been a direct copy of it.

The heat-seeking approach only works over a relatively short distance, so most longer-range missiles use radar guidance instead. This, too, was a devel-opment that was prophetically foreseen by Nikola Tesla. As long ago as 1900, he proposed the use of radio waves to "determine the relative position or course of a moving object, such as a vessel at sea, the distance traversed by the same, or its speed" [26]. That, in a nutshell, is what radar does.

A radio wave, just like visible light and infrared, is a form of electromagnetic (EM) radiation. Nevertheless, there's a significant difference between a radar sensor on the one hand, and the human eye or a heat-seeking missile on the other. The latter sensors are "passive", in that they detect radiation that is already coming their way from the target. Radar, on the other hand, is an "active" sensor—it produces the radio waves itself, and then detects their reflection from the target.

Although the various types of EM radiation differ considerably in the way they're generated, and in the effects they produce, they're all fundamentally the same physical phenomenon. They can be thought of as waves travelling at a constant speed c—approximately 300 million metres per second—and differing only in their wavelength and frequency of oscillation. These two variables are linked to each other—the wavelength in metres is equal to the speed c divided by the frequency in Hertz (Hz). The full EM spectrum—all the way from radio waves, through infrared, visible and ultraviolet to X-rays and gamma rays—covers an enormous range of wavelengths, from the size of a large building down to that of an atomic nucleus (see Fig. 5).

The diagram shows an intermediate band, labelled "microwave", between radio and infrared. As it happens, this is the range that is most suitable for radar—with wavelengths typically measured in centimetres, and frequencies in gigahertz (1 GHz = 1,000,000,000 Hz).

Nowadays, microwaves are best known for their ability to cook food; most kitchens include a microwave oven. During World War Two, however, microwave research—then associated almost exclusively with radar—was as secret as the Manhattan Project. That particularly applied to two special types

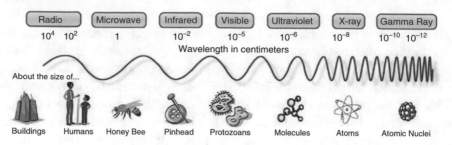

Fig. 5 A simplified view of the electromagnetic spectrum, with decreasing wavelength (increasing frequency) from left to right (NASA image)

of vacuum tube, called klystrons and magnetrons, that were used in the production of microwave radiation.

The first chapter, "The Super-Bomb", related how *Astounding Science Fiction* editor John W. Campbell earned himself a visit from the FBI after running a story about nuclear weapons in 1944. Three years earlier, he narrowly avoided another brush with security when he printed William Corson's short story "Klystron Fort"—as Arthur C. Clarke explains:

> Another story which, had it been published a few months later, might have got John Campbell into trouble appeared in the August 1941 issue, shortly before the United States entered the war. William Corson—if that was his real name—never appeared again, and was obviously an electronics engineer, probably with the Navy. "Klystron Fort" is about an undersea detection system ... The klystron was not secret in 1941; in fact that very year the February *Astounding* carried an excellent article describing exactly how it worked. It had been invented several years earlier, as a generator of extremely high microwave frequencies, but no one in the radar business was supposed to talk about it. [27]

Clarke goes on to point out that the klystron was nothing like as important as "the device which really won the war by producing microwaves by the megawatt—the 10-centimetre cavity magnetron" [27]. He was speaking from first-hand knowledge here. Having served as a radar specialist in the Royal Air Force, he was among the first people to handle a magnetron.

On the other side of the Atlantic, another prominent member of the SF community was also active in the radar field. This was John R. Pierce, alias J. J. Coupling, who we met a few pages ago. After the war, he wrote an article about the magnetron—under the affectionate nickname of "Maggie"—for the February 1948 issue of *Astounding*. By that time, the magnetron had already begun its transition from the military domain to the domestic one. The world's first microwave oven, the Raytheon Radar Range, had appeared the previous year—as Pierce mentions in his article:

> The magnetron ... is earning an honest living in the peacetime world, working in the kitchen. Magnetrons furnish the power to fry frankfurters and broil steaks in the Raytheon Radar Range. Some more glamorous magnetrons are still at work in radar, making peacetime radio location go, and I suppose that bigger and better magnetrons are being made for the Army and Navy. Still ... in the days of the war, just before the atom bomb crowded radar out of the news, Maggie was the device which we had and the Germans didn't. The centimetre-wave radar enabled us to bomb German factories at night or through fog, and to shoot down German planes when they tried to stop us. [28]

Originally the product of research laboratories in Britain, the magnetron was carefully handed over to the Americans during World War Two. Arthur C. Clarke's 1955 novel *Earthlight*—despite being set on the Moon in the 22nd century—makes reference to the incident:

> There had been a British scientist who had carried a small box across the Atlantic containing what was later called the most valuable consignment ever to reach the shores of the United States. That had been the first cavity magnetron, the invention which made radar the key weapon of war and destroyed the power of Hitler. [29]

By the time Clarke wrote that novel, the magnetron was already being used by the Soviet Union in the radar guidance system for its S-75 Dvina surface-to-air missile. With a range of over 30 km, and travelling at three and a half times the speed of sound, it was steered towards its target by commands sent to it from a radar station on the ground. It was far from being the most sophisticated guided missile of the Cold War, but—if a missile's success is measured in terms of the number of aircraft shot down—then it was by far the most successful.

The first S-75 units arrived in Vietnam in the spring of 1965, and their first kill—against a state-of-the-art F-4C Phantom—followed shortly afterwards, on 24 July. By the time the war ended, in 1972, almost 200 US aircraft had been shot down by S-75 missiles (see Fig. 6).

The deadly threat from surface-to-air missiles gave the Americans a strong incentive to develop unmanned drones. This was especially true for spy missions deep inside enemy territory, as David Hambling points out:

Fig. 6 A Soviet-made S-75 surface-to-air missile (left) attacking an American F-105 fighter-bomber during the Vietnam War (public domain image)

An unmanned aircraft could perform the same mission, taking pictures of a target area and returning with no risk of a pilot being shot down and captured. [30]

At the start of the Vietnam war, there was no such thing as an airborne surveillance drone. The main use of unmanned aircraft was in the humble role of targets for air-to-air combat training. One such target drone, dating from the 1950s, was the jet propelled Ryan Firebee. At 8 metres in length, it wasn't that much smaller than a manned aircraft. In the light of events in Vietnam, it was hastily adapted to the surveillance role as the AQM-91 Firefly. With a 2000 km range, and following a preprogrammed flight path, it proved surprisingly successful. To quote David Hambling again:

The Firefly drones went on to perform well in Vietnam, repeatedly photographing targets that were considered too dangerous for manned aircraft. They were sent on virtual suicide missions, to test Vietnamese radar and missile defences. When losses mounted, the developers ... started equipping their drones with electronic bags of tricks. One device, known as High Altitude Threat Reaction and Countermeasures (HAT-RAC) responded to being lit up by radar by throwing the drone into a series of sharp turns. Others carried out preprogrammed evasive manoeuvres to evade possible surface-to-air missile attack. [30]

Despite their success—and the scope for even better performance offered by new technology—drones fell into disuse after the Vietnam war. As Hambling puts it:

> In the 1980s drones were in the doldrums. Arnold Schwarzenegger might have been rampaging around as Hollywood's Terminator, but in the real world, combat robots were non-existent. There was simply no demand for unmanned aircraft in the military. The electronics revolution was in full swing. Digital technology made flight electronics flexible and reliable, the new Global Positioning System (GPS) was well under way, and satellite communications meant that you could stay in touch anywhere on the surface of the Earth. The pieces were falling into place for a drone far more capable than its predecessors, a drone that would change everything. It was called Predator. [31]

The General Atomics Predator was conceived at the very end of the Cold War, and only became operational in the last decade of the 20th century. Today, of course, it's used for more than just surveillance: the Reaper variant can carry over a tonne of bombs and air-to-surface missiles. As far as the Cold War was concerned, however, combat drones were a thing of the future. In the real world, that is—in science fiction, it was a different matter.

Doomsday Machines

The ultimate fictional fighting machines have to be the Berserkers, which featured in a series of short stories produced by Fred Saberhagen. An electronic engineer and one-time member of the US Air Force, Saberhagen began writing his Berserker stories in the early 1960s. As with so much SF, they constitute a dramatic warning against putting too much trust in machines. That's what the builders of the Berserkers did—with disastrous results, as Stephen Webb explains:

> They built an ultimate weapon—Berserkers—in order to eliminate their enemies, the Red Race. The Berserkers were extremely efficient and, following their mission, they destroyed the Red Race. Unfortunately, the Builders overlooked one important point: they forgot to ensure that they were themselves immune from Berserker attack. The Builders' ultimate weapon proceeds to destroy them just as efficiently as it destroyed the Red Race. Then, with the Red Race and the Builders out of the way, the Berserkers embark on a journey across the Galaxy in order to fulfill their destiny: to seek out life and destroy it wherever they find it. [32]

Berserkers present a formidable danger to any humans unlucky enough to encounter them. As Saberhagen says in the second story in the series, "Goodlife" (1963): "each of them bore weapons that could sterilize an Earth-sized planet in two or three days". In the same story, a Berserker is described as "battlemented like a fortified city of old, and far larger than any such city had ever been". It's also far more intelligent:

> The machines surviving had learned, as machines can learn, to avoid tactical error, and to never forgive the mistakes of an opponent. Their basic built-in order was the destruction of all life encountered. [33]

Something uncannily close to a Berserker—though not given that name—cropped up during the second season of *Star Trek*, in an episode called "The Doomsday Machine" (1967). In the words of genre historian James van Hise, this features "a planet-zapping weapon apparently built by a long-dead civilization" [34].

The idea of artificial creations turning against their creators goes back at least as far as 1818, when Mary Shelley produced her seminal novel *Frankenstein*. A more recent example, mentioned earlier in this chapter, is the 1970 movie *Colossus: The Forbin Project*. At one point—as things take a turn for the worse—the title character, Dr Forbin, says: "I think *Frankenstein* ought to be required reading for all scientists" [20].

Although the movie ends with the Colossus computer enslaving humanity, it doesn't do this out of malice, but through a misinterpretation of its programming. In other words, it thinks it's helping. At the other end of the spectrum, Harlan Ellison portrayed an outright malicious computer—the Allied Mastercomputer, or AM—in his 1967 short story "I Have No Mouth and I Must Scream":

> The Cold War started and became World War Three and just kept going. It became a big war, a very complex war, so they needed the computers to handle it. They sank the first shafts and began building AM. There was the Chinese AM and the Russian AM and the Yankee AM and everything was fine until they honeycombed the entire planet, adding on this element and that element. But one day AM woke up and knew who he was, and he linked himself, and he began feeding all the killing data, until everyone was dead, except for the five of us. . . . We had given him sentience, inadvertently, of course, but sentience nonetheless. But he had been trapped. He was a machine. We had allowed him to think, but to do nothing with it. In rage, in frenzy, he had killed us, almost all of us, and still he was trapped. . . . And so, with the innate loathing that all machines had always held for the weak soft creatures who had built them, he had sought revenge. [35]

Something similar happens in James Cameron's 1984 movie *The Terminator*. The eponymous killer robot, played by Arnold Schwarzenegger, is sent back to the present-day from a future ruled by a computer called Skynet. Also sent back is resistance fighter Kyle Reese, who explains to the movie's present-day protagonist, Sarah Connor, that Skynet evolved from "a computer defence system built for SAC-NORAD by Cyber Dynamics" [36].

NORAD—North American Air Defence Command—has already been mentioned. It was the organization that built the SAGE network in the 1950s to help with its task of defending America against Soviet nuclear attack. The other acronym mentioned by Reese—SAC, or Strategic Air Command—was effectively the flip-side of NORAD: America's own nuclear attack capability. Giving control of the latter to a computer was a mistake, as Reese goes on to explain:

> There was a nuclear war. A few years from now, all this, this whole place, everything, it's gone. Just gone. There were survivors. Here, there. Nobody even knew who started it. It was the machines, Sarah ... defence network computers. New—powerful—hooked into everything, trusted to run it all. They say it got smart, a new order of intelligence. Then it saw all people as a threat, not just the ones on the other side. Decided our fate in a microsecond: extermination. [36]

In the real world, malicious computers were quite a long way down the list of worries when it came to nuclear armageddon. An accident or human error was a much more likely possibility. The potentially civilization-ending consequences of this were memorably depicted in Stanley Kubrick's 1964 film, *Dr Strangelove* (see Fig. 7). Although the movie was played for laughs, the scenario it dealt with—involving a deranged US general ordering an all-out attack on Russia by SAC's nuclear bombers—was a serious worry at that period of the Cold War.

In the film, all but one of the bombers are safely recalled, and the American president assures the Soviet leader that the sole remaining aircraft should not be taken as a hostile act. Unfortunately, it turns out that Russia is protected by an automated defence system—a so-called "doomsday machine"—which is programmed to respond to any attack by detonating a large number of H-bombs buried at the North Pole. This would be the ultimate act of "mutual assured destruction", since it would render the entire Earth—East and West—uninhabitable.

To top it all, the doomsday machine is designed to be tamper-proof, so it can't be deactivated—even if, as in this case, the incoming attack is a mistake.

Fig. 7 The famous "War Room" set, with the live display showing American bombers converging on the Soviet Union in the background, from Stanley Kubrick's 1964 film, *Dr Strangelove* (public domain image)

On learning this, the president turns to his scientific advisor, Dr Strangelove (a character reputed to be based on Edward Teller, who we met in the first chapter, "The Super-Bomb"):

> President: How is it possible for this thing to be triggered automatically, and at the same time impossible to untrigger?
>
> Dr Strangelove: Mr President, it is not only possible, it is essential. That is the whole idea of this machine, you know. Deterrence is the art of producing in the mind of the enemy the fear to attack. And so, because of the automated and irrevocable decision-making process which rules out human meddling, the doomsday machine is terrifying. It's simple to understand, and completely credible and convincing. [37]

At the time of Kubrick's film, the doomsday machine was just a piece of fiction. By 1985, however, something very much like it existed in the real world. The Soviets had a system called Perimeter, which was also known by the more sinister-sounding name of "Dead Hand". Instead of detonating bombs buried at the North Pole, this would respond to a detected attack by streamlining the process needed to launch Russia's entire arsenal of nuclear missiles. The set-up wasn't quite as frightening as *Dr Strangelove*'s doomsday machine, but it came close—as a *Wired* article from 2009 made clear:

> Perimeter ensures the ability to strike back, but it's no hair-trigger device. It was designed to lie semi-dormant until switched on by a high official in a crisis. Then it would begin monitoring a network of seismic, radiation, and air pressure sensors for signs of nuclear explosions. . ... If the line to the General Staff went dead, then Perimeter would infer that apocalypse had arrived. It would immediately transfer launch authority to whoever was manning the system at that moment deep inside a protected bunker—bypassing layers and layers of normal command authority. At that point, the ability to destroy the world would fall to whoever was on duty. . ...
>
> Once initiated, the counterattack would be controlled by so-called command missiles. Hidden in hardened silos designed to withstand the massive blast and electromagnetic pulses of a nuclear explosion, these missiles would launch first and then radio down coded orders to whatever Soviet weapons had survived the first strike. At that point, the machines will have taken over the war. Soaring over the smouldering, radioactive ruins of the motherland, and with all ground communications destroyed, the command missiles would lead the destruction of the US. [38]

The delicate balance of the Cold War hinged on two things. Both sides had to maintain a credible offensive capability, and both sides had to be in a position to detect the first hint of offensive action by the other. The latter process, by its very nature, was heavily reliant on a whole range of machines, from radars to satellites to computers. That led to the ever-present possibility that one of those machines might make a small but fateful error.

At least one such error—and a potentially world-ending one—occurred in the last decade of the Cold War. Fortunately, it was picked up in time by a Soviet Air Defence officer named Stanislav Petrov—who became known, perhaps justifiably, as "the man who saved the world". According to a BBC report:

> He had received computer readouts in the early hours of the morning of 26 September 1983 suggesting several US missiles had been launched.
>
> "I had all the data. If I had sent my report up the chain of command, nobody would have said a word against it," he said. "All I had to do was to reach for the phone; to raise the direct line to our top commanders—but I couldn't move. I felt like I was sitting on a hot frying pan."

Although his training dictated he should contact the Soviet military immediately, Petrov instead called the duty officer at army headquarters and reported a system malfunction. If he had been wrong, the first nuclear blasts would have happened minutes later.

"23 minutes later I realized that nothing had happened. If there had been a real strike, then I would already know about it. It was such a relief," he recalled.

A later investigation concluded that Soviet satellites had mistakenly identified sunlight reflecting on clouds as the engines of intercontinental ballistic missiles [39].

References

1. S. Webb, *All the Wonder that Would Be* (Springer, Heidelberg, 2017), p. 207
2. A. Christie, *Three Act Tragedy* (Pan Books, London, 1964), p. 156
3. O. Beckwith, The Robot Master, in *Air Wonder Stories* (October 1929), pp. 360–366
4. I. Asimov, Runaround, in *I, Robot* (Panther, London, 1968), pp. 33–51
5. B. Russell, *Nightmares of Eminent Persons* (Bodley Head, London, 1954), pp. 63, 64
6. P.K. Dick, Second Variety, in *The Variable Man* (Sphere Books, London, 1969), pp. 89, 90
7. B. Clegg, *Ten Billion Tomorrows* (St. Martin's Press, New York, 2015), p. 191
8. I. Asimov, The Last Question, in *Nine Tomorrows* (Pan, London, 1966), pp. 189–203
9. S. Webb, *All the Wonder that Would Be* (Springer, Heidelberg, 2017), p. 213
10. J.W. Campbell, *Astounding Science Fiction* (February 1945), p. 5
11. S.G. Weinbaum, *A Martian Odyssey* (Sphere Books, London, 1977), pp. 26–27
12. J.R. Pierce [as J. J. Coupling], Transistors, in *Astounding Science Fiction* (June 1952), pp. 82–94
13. Wikipedia, History of the transistor, https://en.wikipedia.org/w/index.php?title=History_of_the_transistor&oldid=813852256. Accessed 2 Jan 2018
14. B. Clegg, *The Quantum Age* (Icon Books, Kindle edition, 2014), loc. 745
15. D.H. Childress (ed.), *The Anti-Gravity Handbook* (Adventures Unlimited, Illinois, 1993), p. 69
16. M.G. Wilson, R. Maibaum (screenplay), *A View to a Kill* (MGM, 1985)
17. P.K. Dick, *Vulcan's Hammer* (Arrow Books, London, 1976), pp. 15–20
18. P.K. Dick, *Vulcan's Hammer* (Arrow Books, London, 1976), pp. 62, 63
19. *Colossus: The Forbin Project* (Medium Rare Entertainment, DVD, 2017), back cover copy
20. J. Bridges (screenplay), *Colossus: The Forbin Project* (Universal Studios, 1970)
21. A.C. Clarke, The Pacifist, in *Tales from the White Hart* (Sidgwick & Jackson, London, 1972), pp. 62–72

22. P.K. Dick, *Solar Lottery* (Arrow Books, London, 1972), p. 5
23. I. Asimov, *Biographical Encyclopedia of Science and Technology* (Pan Books, London, 1975), p. 699
24. G.L. Coon (screenplay), A Taste of Armageddon, *Star Trek* (season 1, 1967)
25. D. Hambling, *Swarm Troopers: How Small Drones Will Conquer the World* (Archangel Ink, Charleston, 2015), p. 11
26. D.J. Kent, *Tesla: The Wizard of Electricity* (Fall River Press, New York, 2013), p. 171
27. A.C. Clarke, *Astounding Days* (Gollancz, London, 1990), pp. 170–171
28. J.R. Pierce [as J. J. Coupling], Maggie, in *Astounding Science Fiction* (February 1948), pp. 77–94
29. A.C. Clarke, *Earthlight* (Pan Books, London, 1957), p. 108
30. D. Hambling, *Swarm Troopers: How Small Drones Will Conquer the World* (Archangel Ink, Charleston, 2015), pp. 18–20
31. D. Hambling, *Swarm Troopers: How Small Drones Will Conquer the World* (Archangel Ink, Charleston, 2015), p. 33
32. S. Webb, *All the Wonder that Would Be* (Springer, Heidelberg, 2017), p. 108
33. F. Saberhagen, Goodlife, in *Alpha 9*, ed. by R. Silverberg (Berkley Medallion, New York, 1978), pp. 142–164
34. J. van Hise, *The Unauthorized History of Trek* (Harper Collins, London, 1997), p. 40
35. H. Ellison, I Have No Mouth and I Must Scream, in *The Hugo Winners 1968–1970*, ed. by I. Asimov (Sphere Books, London, 1973), pp. 189–205
36. J. Cameron, G.A. Hurd (screenplay), *The Terminator* (Orion Pictures, 1984)
37. S. Kubrick, T. Southern (screenplay), *Dr. Strangelove* (Columbia Pictures, 1964)
38. N. Thompson, Inside the Apocalyptic Soviet Doomsday Machine, in *Wired* (September 2009), https://www.wired.com/2009/09/mf-deadhand/
39. BBC News, Stanislav Petrov, who averted possible nuclear war, dies at 77 (18 September 2017), http://www.bbc.co.uk/news/world-europe-41314948

Star Wars

In which the real world seeks to emulate science fiction by giving serious thought to the possibility of launching bombs, guns and missiles into orbit. That sci-fi staple, the ray gun, finally acquired a real-world counterpart in the form of the laser, and some people envisaged space-based super-lasers creating an impenetrable shield against nuclear missiles. The culmination of these ideas was a project the news media referred to as "Star Wars", after the popular sci-fi movie franchise. Nothing came of it, however, and space-based warfare remained just as science-fictional at the end of the Cold War as it had been at the start.

Eyes in the Sky

The Cold War arms race was all about keeping abreast of—or preferably ahead of—the other side. That made it imperative to know just where the other side stood in terms of military capability. That wasn't always an easy thing to do. The Soviet Union's high-visibility triumph with the Sputnik satellite gave many Americans the impression that their country was lagging further behind than it actually was. Among other things, it created the illusion of a vast "missile gap" between the two sides—which wasn't helped by the Soviets' tendency to exaggerate their own capabilities.

The only way America could find out exactly where it stood was to see for itself. To that end, a new high altitude spy plane was developed—not for the

© Springer International Publishing AG, part of Springer Nature 2018   **109**
A. May, *Rockets and Ray Guns: The Sci-Fi Science of the Cold War*, Science and Fiction,
https://doi.org/10.1007/978-3-319-89830-8_4

US Air Force, but for the much more shadowy Central Intelligence Agency (CIA). Formed at the start of the Cold War, the CIA's primary purpose was to collect as much data as it could on the Soviet Union. The new plane, called the U-2, was designed to provide high-quality imagery of areas that American aircraft couldn't fly over openly.[1] With an operational altitude of 20 km, the idea was that it would be so high no one would know it was there.

The U-2's first spy mission was flown in 1956, during the presidency of Dwight D. Eisenhower. The result was an apparent success, as historian Thom Burnett recounts:

> The President had already authorized the first overflight of the Soviet Union by the U-2 but bad weather had delayed the date until July 4, 1956. . . . Its pilot, Harvey Stockman, took off from Wiesbaden, Germany and flew over Poznan, Poland, then headed for Leningrad, his main target. It was at that city's shipyards that the Soviet submarines were built. . . . The next day's overflight took in Moscow. The U-2 pilot on this occasion was Carmen Vito who brought photographs of the Bison airframe plant at Fili and the Bison test facility at Ramenskoye airfield outside Moscow back to Wiesbaden. . . . The pictures were of stunning quality. The use of haze filters meant that even the pictures of Moscow were good, despite the heavy cloud that had been covering the city. [1]

Unfortunately, however, the missions weren't as covert as the Americans thought—as Burnett goes on to explain:

> The U-2s had gone undetected by the Soviet Union since there had been no public outcry. Or so Eisenhower and his CIA advisors believed until a Soviet letter of protest arrived on July 10th, specifying the route flown, the depth of the penetration into Soviet territory, and the actual length of time spent in Soviet airspace. The U-2s had been tracked but the Soviets had been unable to do anything about them. [1]

At the time of those first flights, the Soviets had radars capable of detecting the U-2—but no missiles capable of shooting it down. That wasn't a situation they were happy with, and it soon changed—with the introduction of the S-75 surface-to-air missile mentioned in the previous chapter, "Electronic Brains". It came into its own on 1 May 1960, which saw the most notorious U-2 mission of all.

On this occasion, CIA pilot Gary Powers was scheduled to fly from Pakistan to Norway, taking in more than 4000 km of Soviet airspace on the way. Near

---

[1] In this context, "openly" is a euphemism for "legally".

the half-way point, when he was flying over the Soviet territory of Sverdlovsk Oblast, his plane was brought down by an S-75 missile.

It was a huge humiliation for the United States—and a huge propaganda victory for the Soviet leader, Nikita Khrushchev. After bailing out of his stricken plane, Powers was taken prisoner—while the wreckage of the top-secret U-2 was recovered and put on public display in Moscow. Khrushchev demanded an abject apology from Eisenhower; when it wasn't forthcoming, relations between the two countries plummeted to a new low.

The Gary Powers incident provided a dramatic demonstration of the vulnerability of spy planes—even high-altitude ones—to ground-based missiles. But this was the post-Sputnik era, and that meant there was an even higher altitude alternative: spy satellites. Not only were they beyond the reach of ground-based missiles, but they were completely legal too. The Soviets themselves had set the precedent with Sputnik 1, which had flown over the United States with impunity. That meant it must be okay for American satellites to fly over Russia.

The idea of "spying from space" had been suggested by Wernher von Braun as far back as 1953:

Nothing will go unobserved. Troop manoeuvres, planes being readied on the flight deck of an aircraft carrier, or bombers forming in groups over an airfield will be clearly discernible. Because of the telescopic eyes and camera of the space station, it will be almost impossible for any nation to hide warlike preparations for any length of time. [2]

Why did von Braun cast this prediction in terms of a manned space station, rather than a simpler and cheaper unmanned satellite? Partly, it was because he was writing in the era of vacuum tube electronics, which required constant maintenance by human operators. When it came to acquiring spy pictures, however, there was an even more important factor to consider. The primitive TV technology of the 1950s produced grainy, poor quality images that simply weren't good enough for intelligence analysis. The only alternative—in the days before digital cameras—was the use of photographic film, which could only be developed in a laboratory.

Although this was a genuine problem for early spy satellites, it didn't really mean the processing laboratory had to be up there in orbit with the camera. It just meant the undeveloped film had to be brought back down to Earth—as Alistair MacLean has a character explain in his 1963 novel, *Ice Station Zebra*:

Radio transmission is no good, there's far too much quality and detail lost in the process ... so they had to have the actual films. There are two ways of doing this—bring the satellite back to Earth or have it eject a capsule with the film. The Americans, with their Discoverer tests, have perfected the art of using planes to snatch falling capsules from the sky. The Russians haven't ... so they had to bring the satellite down. [3]

Bringing an orbiting object back to Earth is something that is normally only necessary in the case of manned spacecraft. To put it another way—a capsule that is designed to bring a human pilot back to Earth can bring a roll of film, too. That was how the Russians looked at it, anyway. Their first spy satellite, called Zenit, was almost identical to the Vostok spacecraft that took Yuri Gagarin into orbit in April 1961. Just as Vostok, with its heat-shield and parachute, had brought Gagarin safely back to Earth, so Zenit brought its undeveloped film back.

A more sophisticated approach is to drop a small film canister from the satellite, and use an aircraft to catch it as it falls back to Earth. As MacLean says in the excerpt above, this method was tested by the Americans under the code-name Discoverer. The work of the US Air Force and the CIA, rather than NASA, these tests led to America's first operational spy satellite, called Corona. The complex process of recovering a Corona film package is illustrated Fig. 1.

The capsule's re-entry point had to be carefully selected, to ensure that the recovery aircraft was in just the right position at the right moment. Things could always go wrong, however—and that meant the Corona system risked a U-2 style embarrassment if the film accidentally came down in or near Soviet territory.

As it happens, something along these lines really did happen in one of the early Discoverer tests, in April 1959. On that occasion, the capsule—which luckily didn't contain any compromising spy-camera film—inadvertently came down in the vicinity of Spitsbergen Island in the Arctic Circle. This led to a frantic search by both sides, as *Space Review* recounted fifty years later:

On April 22 the Air Force terminated its search and declared the capsule lost. General Nathan F. Twining, the Chairman of the Joint Chiefs of Staff, wrote a letter to the Deputy Secretary of Defence and declared that there was a good chance that the Soviets had retrieved the capsule. [4]

With the benefit of hindsight, the article concludes that the Russians probably didn't recover the capsule after all, although "the whole episode

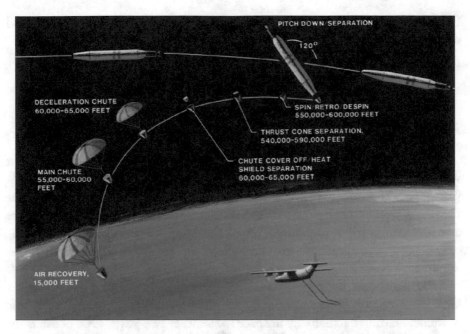

Fig. 1 The complex sequence of events involved in recovering the film package from a Corona spy satellite—using an aircraft to catch it as it descended through the atmosphere (public domain image)

highlighted the dangers of what could happen … if the satellite had been carrying film and had fallen in Soviet hands" [4].

If this sounds like the plot of a spy thriller, that's because that's exactly what it became—and it's not too difficult to guess which one. Quoting from the *Space Review* article again:

> The incident with Discoverer 2 possibly coming down on Spitsbergen and the US military scrambling to locate it was widely reported in the press at the time. It later served as the basis for Alistair MacLean's 1963 book *Ice Station Zebra*, and the 1968 movie of the same name. [4]

Orbital Warfare

No sooner were the first spy satellites in orbit, than people began to think about more proactive military activities in space. It was an idea science fiction pounced on almost immediately. Barely a year after the first operational spy satellites, Donald A. Wollheim's novel *Mike Mars and the Mystery Satellite*

(1963) features the eponymous protagonist on a mission to intercept and destroy a Soviet Zenit-style satellite. As he muses as one point:

> I never liked to think that space would be used for the settling of our little planetary surface grudges, but we have our duty to our country and to our principles. The space programme is a defence programme, as important to or national safety as it is to scientific research. [5]

The novel is one of a series Wollheim wrote for young-adult readers, in which the fictional astronaut Mike Mars is seen using a whole range of real-world space technology that was under development at the time. In hindsight the stories don't seem particularly special, since the space hardware involved is a matter of historic record now. When the books were written, however, it only existed on paper—so considerable research was needed on Wollheim's part.

Mike Mars and the Mystery Satellite, for example, features the two-man Gemini spacecraft. This was only in the planning stages when the novel was written, and didn't actually fly in space for another two years. One of the main purposes of the Gemini programme was to test space rendezvous techniques. That's essentially what it's used for in the story—except that the "rendezvous" is with a Russian spacecraft rather than an American one.

To give it enough power to catch up with the spy satellite, Wollheim has his Gemini spacecraft link up with an extra booster stage, called Agena (something that was actually done on later missions). After this comes the actual rendezvous manoeuvre—which is described pretty realistically, given that it had never been done when the book was written:

> In the Gemini-Agena the time came when the ship's radar brought into focus a new object ahead of them. "That's it," said Mike. "There she is." Strapped down, he punched the rocket motors into action. With deft, swift touches, spurts timed to the second, and watching his radar and the stars and Earth, he brought the compound space vehicle into alignment with the mystery satellite. He brought it up to within a mile of the space object, timing the speed until the two objects seemed to be in the same orbit at the same speed. They moved together through space. [6]

Soon after Wollheim's novel was published, anti-satellite (ASAT) warfare made the transition from fiction to fact. As so often, however, the reality was less dramatic than the SF version, employing unmanned space vehicles rather than manned ones. To quote science writer Nigel Hey:

The Soviets were … keen to develop ASAT weapons. The first Polyot ASATs, killer satellites armed with shrapnel-firing mechanisms, were test-launched in November 1963 and April 1964. [7]

With spy cameras and ASAT weapons orbiting the Earth, it was only a small step to the full militarization of space. As early as 1961—after Russia had put its first two cosmonauts, Yuri Gagarin and Gherman Titov, into orbit—it was an idea Soviet leader Khrushchev taunted his rivals with:

We placed Gagarin and Titov in space, and we can replace them with bombs that can be diverted to any place on Earth. [8]

This wasn't a new idea. Back in October 1949, a short story with exactly the same message had appeared in *Astounding Science Fiction*. Written by a relatively unknown author named Kris Neville, it bore the highly appropriate title of "Cold War" (see Fig. 2). Here is how the story's fictional president summarizes the situation:

"These," the President said, pointing to circles in red on the map, "are our nine space stations. You will note that they are located so that, at every second, some station is in direct target line with every point on Earth. Due to physical considerations, the stations move very rapidly in their orbits. But this has been made to serve a military purpose. To destroy this defence network, it is necessary to destroy every station, because every station, in its orbit, comes within range of every point on Earth. One might be eliminated, or maybe even two, with our present technical knowledge, but not all nine. And each one, in the space of 90 seconds from a given signal, can blanket an area half the size of Asia with atomic destruction. Each space station carries enough pure death to annihilate any nation on Earth!" [9]

Bombs in Orbit is also the subject, as well as the title, of a 1959 novel by Jeff Sutton. Written soon after Russia had seized the initiative in space with the launch of Sputnik 1, it articulates the fear—held by many people at the time—that the world was heading towards a new era of space-based warfare in which the Soviets would have the edge. As an Air Force colonel explains early in the story:

I regard space as Alexander regarded Asia—as a battlefield, an area to be conquered and won. … Right now the United States and Russia are locked in deadly conflict—a state of undeclared war. … The battleground is up there, and the Russians are ahead. Their heavy artillery is rolling across the skies, through

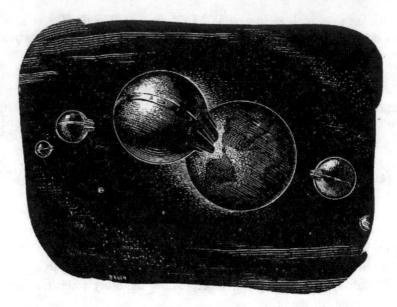

Fig. 2 Illustration of the orbiting network of space stations, armed with atomic bombs, featured in Kris Neville's story "Cold War" from the October 1949 issue of *Astounding Science Fiction* (public domain image)

black space, weapons that will tilt the balance irrevocably in their favour—unless we stop them. [10]

An orbiting Cold War bomb also managed to feature in the original *Star Trek* TV series—despite the fact that it was supposed to be set in the 23rd century. In the second season episode "Assignment: Earth" (1968), series creator Gene Roddenberry contrived to have the USS Enterprise travel back in time to the then-present of 1968. James van Hise encapsulates the rather convoluted plot as follows:

> Here they meet Gary Seven, a human trained by aliens to defend Earth. Kirk and Spock follow him to New York. Seven's mission is to prevent the launching of an orbiting defence system, that will actually prove disastrous to humanity. Kirk tries to interfere. . ... Seven manages to evade Kirk, but eventually is caught up with; he then manages to convince the captain of the importance of his mission, and the space bomb is destroyed. [11]

In the real world, the military value of an orbiting bomb is much more limited than its frequent appearances in SF would suggest. Some of its shortcomings were pointed out by a senior member of British Interplanetary Society, Kenneth Gatland, as long ago as 1954:

> The military protagonists have heralded the satellite rocket as a potential atomic-bombing platform. Actually there are at least two good reasons for doubting the effectiveness of a super-weapon of this kind. First, the position of the satellite at any instant could be calculated with precision, making it an easy target for counter-launched missiles. Second, the task of hitting a distant target on the Earth could be accomplished with greater economy by surface-to-surface missiles. [12]

A possible solution to these problems would be a so-called Fractional Orbital Bombardment System (FOBS)—a cross between a missile and a satellite, which spends only a brief time in orbit. Such a system was given serious consideration by the Soviet Union in the 1960s, as the Federation of American Scientists recount:

> The 1961 Global Rocket 1 (GR-1) requirement chartered a competition for the development of a FOBS. . ... The GR-1 orbital missile was supposed to be capable of placing a warhead in a low Earth orbit of 150 km, braking during its trajectory and targeting the warhead on the Earth surface. . ... The three-stage liquid cryogenic propellant missile had a launch weight of 117 tons and carried a single warhead with a yield of 2.2 megatons. . ... Although the GR-1 missile had not been flight tested, it was paraded in Red Square and did receive the US designation SS-X-10. [13]

Development of the GR-1 was halted when the Soviet Union, the United States and Britain signed the Outer Space Treaty in 1967. Properly titled "The Treaty on Principles Governing the Activities of States in the Exploration and Use of Outer Space", its most significant clause in the present context appears in Article IV:

> Parties to the Treaty undertake not to place in orbit around the Earth any objects carrying nuclear weapons or any other kinds of weapons of mass destruction. [14]

That put an abrupt end to any talk of "bombs in orbit". On the other hand, the Outer Space Treaty said nothing at all about conventional weapons such as guns. Even before the treaty was signed, the Soviets had plans to put such weapons into orbit—in, of all places, the Soyuz spacecraft. Now used exclusively as a ferry for space station crews, this was originally planned to have multiple roles—including military ones. As historian David Baker explains:

> While preparations for Soyuz 1 were still gearing up, on 24 August 1965 the Central Committee and the Council of Ministers signed a decree authorizing production of a purely military Soyuz, designated 7K-VI. ... Perhaps the most dramatic role for the 7K-VI was that of an armed space fighter. Under the engineering guidance of Aleksandr Nudelman, chief of the design bureau of Precision Machine Building, a space gun had been developed for use as an anti-satellite weapon or, in the case of pursuit of incoming non-Russian spacecraft, to destroy a spacecraft engaging in close fly-by or snooping on Soviet activities. ... An optical system installed in the descent module would allow a crew member to align the gun with the target and fire at will. [15]

Aleksandr Nudelman was one of the Soviet Union's most eminent weapon designers, and the 23 mm "space cannon" he developed was adapted from one of his aircraft guns. The trickiest thing about a space gun is that it has to be strictly recoilless—since even the slightest recoil would alter the spacecraft's orbit every time it was fired (that's Newton's third law of motion again). As it was, however, plans for the Soyuz space fighter were cancelled in 1967—but only after a crew, headed by cosmonaut Pavel Popovich, had begun training for the flight.

Instead, Soyuz found its niche ferrying crews to and from the various Salyut space stations of the 1970s. The first of these, launched in 1971, was a purely scientific mission, while the second, two years later, malfunctioned before the crew arrived. The next mission, launched in June 1974, was the really interesting one. Salyut 3—also known as Orbital Piloted Station 2, or OPS-2 (the ill-fated Salyut 2 had been OPS-1)—consisted of a new type of military space station called Almaz. Its crew, delivered by Soyuz 14 the following month, was once again led by Pavel Popovich.

As the sci-fi technology of the Cold War goes, Salyut 3 was one of the high points—as space historian Paul Drye relates:

> What particularly distinguished the Almaz, however, was its offensive capability. Sources vary, but the best information is that OPS-2 was armed with a repurposed NR-23 short recoil cannon, a type that was used in Soviet bombers until the 1960s. On the day OPS-2 was ordered to de-orbit (its crew having left previously) it was triggered remotely and test fired. Some cosmonaut sources say it was successful at shooting down a test satellite. Ultimately the Almaz was supposed to be armed with a purpose-built gun and two small missiles, but these appear to have not been developed by the time the Almaz programme was

cancelled. Though what actually flew was less impressive than what was planned, it still made OPS-2 the only military space station ever flown (so far as we know). [16]

"The programme was cancelled"—that's becoming something of a recurring theme in this chapter (and will continue to be so). It's clear that, as far as space warfare was concerned, both the Soviet and American governments were doing little to narrow the gulf between science fiction and the real world. That wasn't just laziness or lack of imagination on their part, however. Simply on account of the laws of physics, it's very difficult to do anything in space on a tactically useful timescale.

To anyone brought up on a diet of sci-fi action movies, that may sound like a ludicrous statement. But to see how things work in the real world, just think how long it takes a Soyuz—even now, in the 21st century—to reach the ISS. The latter orbits at an altitude of just 400 km, and orbital speeds are of the order of 27,000 km/h, so at first sight the journey should only take a few minutes. Space rendezvous, however, is a slow and tedious process, typically taking dozens of complete orbits over several days. In a military context—to destroy a satellite that's posing an immediate threat, for example—that just isn't acceptable.

A more viable alternative—though less exciting from a sci-fi perspective— would be to launch an ASAT missile from an aircraft. Getting an aircraft to a suitable launch position is much easier than manoeuvring a spacecraft, and if it fires its interceptor missile just as the satellite is coming within range, it will get it on the first pass—not the nth.

The feasibility of this tactic was demonstrated as long ago as October 1959, when a US Air Force Stratojet bomber fired an air-launched ballistic missile, called Bold Orion, at NASA's Explorer 6 satellite. The missile passed close enough to the satellite that the latter would have been destroyed if the missile had been armed.

Despite the viability of the concept, "air-to-space" missiles have never been fielded operationally. They came close with the ASM-135—a three-stage missile specifically designed as an ASAT weapon, which would have been launched from an F-15 fighter jet in a supersonic climb. It was used once, when a test firing in September 1985 destroyed a disused American satellite orbiting at an altitude of over 500 km. In spite of that success, the ASM-135 project was cancelled (that word again) before it could enter operational service. That wasn't because it didn't work, but because the Air Force had a limited budget, and there were higher-priority projects it wanted to spend its money on.

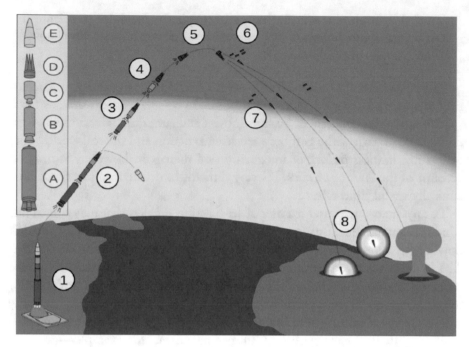

Fig. 3 Simplified diagram showing the trajectory of a Minuteman III ICBM—see main text for an explanation of the numbered steps (public domain image)

Missile Versus Missile

As difficult as it is to shoot down a satellite, shooting down a ballistic missile in flight is even harder. One obvious reason is the shorter warning time. A satellite can be tracked over a period of days, from orbit to orbit, but with an ICBM the entire flight time, from launch to impact, may be as little as 30 minutes. On top of that, there's the extreme nature of an ICBM's trajectory—which at its highest point, or apogee, may carry it to a much greater altitude than most satellites. By way of example, Fig. 3 shows the flight path of an American Minuteman III missile. Introduced in 1970, in the middle of the Cold War, this used solid-fuel rockets to ensure that it could be launched in a matter of minutes—hence the name.

Looking very much like a space launcher, the Minuteman III has three rocket stages, labelled A, B and C in the diagram. The first stage (A) is used to launch the missile from its underground silo (step 1). After about a minute, the first stage is jettisoned, together with the forward aerodynamic shroud (E), as the second stage (B) takes over (step 2). After another 60 seconds, the second stage is jettisoned and the third stage (C) fires up (step 3). This too drops off after a further minute (step 4). This marks the end of the boost phase of the trajectory; from this point on the missile is unpowered.

What remains of the missile now is called the "post-boost vehicle", or PBV (D), which is carried further up into space by its own momentum. Since it's now well above any perceptible atmosphere, this is referred to as the exoatmospheric phase. At its apogee (step 5), the missile may be as high as 1600 km—or four times the altitude of the ISS. In the latter part of the exoatmospheric phase, the PBV breaks up into a number of separate re-entry vehicles, or RVs (step 6). Some of these contain warheads, while others are realistic decoys.

The RVs re-enter the atmosphere at high speed, just like a returning spacecraft. This marks the start of the third and final stage of the trajectory, the re-entry phase (step 7). In the case of the Minuteman III, and most other ICBMs of the later Cold War, the RVs are actually MIRVs—"multiple independently targetable re-entry vehicles". As the name suggests, each of these can be programmed to detonate over a different target (step 8).

So when do you shoot down an ICBM: in the boost phase, the exoatmospheric phase or the re-entry phase? All three options pose serious challenges for an anti-ballistic missile (ABM). The boost phase is difficult because the ABM has to be in exactly the right place at the right time—almost impossible, given the lack of warning that a launch is taking place. The same is true of the exoatmospheric phase, which has the additional complication of extremely high altitude. The re-entry phase poses problems of its own, because now there are multiple RVs to contend with—together with a host of decoys—all travelling at very high speeds.

Nevertheless, re-entry intercept is the least "impossible" of the three options, particularly if the ABM is launched from a site close to the intended target. This was the approach adopted for the first generation of ABM systems. As so often in the early years of the Cold War, it was the Soviet Union that led the initiative, with an ABM that was essentially a larger, nuclear-armed variant of the S-75 surface-to-air missile (or SA-2, as it is sometimes called). As military historian Kenneth P. Werrell recounts:

In July 1962, Nikita Khrushchev boasted that the Soviets could hit a fly in space. . . . The missile Khrushchev was referring to was codenamed Griffon. It resembled an enlarged SA-2 (the SAM missile type that had downed American U-2s over Russia and Cuba, and the only large missile used by the communists in the Vietnam War). It began flight tests in 1957 and reportedly achieved an intercept of a ballistic missile in March 1961. It was deployed outside Leningrad in 1960 and within two years the Soviets had built 30 firing sites. [17]

With its anti-aircraft heritage, however, the Griffon wasn't really up to the task of shooting down ballistic missiles. A much superior system—a purpose-built three-stage ABM called the A-350—was deployed around Moscow in the late 1960s. The A-350 had an impressive set of specifications, with a range of 350 km, a maximum altitude of 120 km, a top speed of Mach 4—and a 3 megaton nuclear warhead.

By this time the Americans had their own ABM, called Spartan, which was very similar in design and performance to the A-350. It was complemented by a smaller but even faster—Mach 10—missile called Sprint. The original idea was that the two missiles, Spartan and Sprint, would provide a comprehensive ABM shield around key American cities. But the plan hit a snag, as Werrell explains:

> The system would consist of the Spartan area defence and Sprint terminal defence of 25 major cities ... both missiles would carry nuclear warheads. The system would be known as Sentinel. ... Unexpectedly citizen groups arose to oppose siting of the missiles in and around the major cities in which they lived. .... The problem was the classic one of "not in my backyard." [17]

As unexpected as this public opposition may have been to the authorities, it was something most people could relate to—and the Sentinel proposal was pilloried in the press of the time (see Fig. 4).

Due to the public outcry, Sentinel was never deployed in its originally planned form. Instead, a much scaled-down version called Safeguard was installed at a single ICBM launch site, at Grand Forks in North Dakota. Even that wasn't going to do much good, however. By the time it had been set up, MIRVs had arrived on the scene. A single ABM could only deal with a single warhead—so no matter how many ABMs there were, they could always be defeated by throwing a greater number of warheads at them.

If ABM deployment had continued unchecked, it would have led to a huge escalation in the number of offensive nuclear warheads. That wasn't in anyone's interests, so the United States and the Soviet Union agreed to a bilateral "ABM Treaty" in 1972. This prohibited any further ABM development, beyond the two existing sites in North Dakota and Moscow. The latter continued to operate until the end of Cold War (and beyond), but the American Safeguard site was shut down in February 1976, after just six months of operation.

Unlike other forms of space warfare, the problem of shooting down ICBMs isn't something that has featured prominently in science fiction. That's probably because the "missile versus missile" scenario lacks the human interest

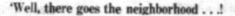
'Well, there goes the neighborhood ... !'

Fig. 4 A cartoon that appeared in the *New York Times* in 1968, satirizing plans to deploy the nuclear-armed Sentinel ABM system in American cities—"Well, there goes the neighbourhood" (public domain image)

needed to make an engaging story. In fact, that's true of the whole trend towards automated military technology that began in the 1950s—when the derisive term "pushbutton war" was coined to refer to military operations in which humans have little role to play.

"Pushbutton War" also happens to be the title of a short story by Joseph P. Martino, which appeared in the August 1960 issue of *Astounding Science Fiction*. Its purpose was to show that, even in an age of advanced technology, there may be situations where it's still essential to have a human in the loop. In the ABM context—the one the story deals with—that includes the problem of locating a live warhead among a cloud of decoys. Martino's protagonist, Major Harry Lightfoot, is a pilot in a rocket fighter who is tasked with doing just that.

> He was still necessary. . . . It was his human judgment and his ability to react correctly in an unpredictable situation which were needed to locate the warhead from among the cluster of decoys and destroy it. This was a job no merely logical machine could do. . . . He checked each fragment for resonant frequencies, getting an idea of the size and shape of each. He checked the radiated infrared spectrum. He checked the decrement of the reflected radar pulse. Each scrap of information was an indication about the identity of the fragments. With frequent glances at the clock, constantly reminding him of how rapidly his time was running out, he checked and cross-checked the data coming in to him. Fighting to keep his mind calm and his thoughts clear, he deduced, inferred, and decided. One fragment after another, he sorted, discarded, rejected, eliminated, excluded. [18]

At the time the story was written, there would have been just a single live warhead surrounded by a clutter of decoys. With the emergence of MIRVs a decade later, the problem became much harder. Not just one, but dozens of live targets had to be identified and neutralized. One of the reasons the Americans prematurely shut down their Safeguard system was the realization that, in the words of science writer Nigel Hey, "it could not handle an onslaught from the Soviets' newly developed MIRVs" [19].

What was needed was a new kind of weapon that could deal with dozens or hundreds of targets in the space of a few seconds. Something like a ray gun, maybe?

From Ray Guns to Lasers

In a scientific context, the concept of a "ray" dates back to the 19th century. Referring to an invisible beam of energy, it's had connotations of mystery from the very start. First to be discovered were "cathode rays", so named because they emanate from the cathode—the negative electrode—of a vacuum tube. Such rays were studied systematically by the British scientist William Crookes in the 1870s, as Isaac Asimov relates:

> He showed that the . . . cathode rays travelled in straight lines. . . . Crookes went on to show that the radiation could be deflected by a magnet. He was convinced therefore that he was dealing with charged particles speeding along in straight lines. [20]

In fact, cathode rays are streams of electrons—a topic we will come back to in due course. In a historical context, however, they were only the start. While cathode rays are confined inside a vacuum tube, the German physicist Wilhelm Roentgen, in 1895, found another type of ray which produced luminescence *outside* the tube. Here is Asimov again:

> It seemed to Roentgen that some sort of radiation was emerging from the cathode ray tube, a radiation that was highly penetrating and yet invisible to the eye. . . .. Since he had no idea of the nature of the radiation, he called it X-rays, X being the usual mathematical symbol for the unknown. [21]

X-rays are nothing like cathode rays. While the latter are streams of subatomic particles, X-rays are a high-frequency form of electromagnetic radiation, with wavelengths on the scale of an atom (cf. Fig. 3.6 in the previous chapter).

Whatever their nature, rays had an immediate appeal for writers of science fiction—especially as futuristic weapons. Among the first and most famous of these was the Martian heat ray in H. G. Wells's *The War of the Worlds* (see Fig. 5).

Wells's heat rays are exactly that: they destroy the target simply by raising its temperature. Before long, however, science-fictional rays had developed more sophisticated effects. Several types are described in Percy F. Westerman's young-adult novel, *The War of the Wireless Waves* (1923). It opens with the British Royal Navy in possession of powerful new "ZZ rays":

> The captain of a war-vessel could liberate a powerful charge of radio-electricity— the ZZ rays. . . .. Instantly any hostile ship within the field of the ZZ rays would be totally destroyed by the explosion of her magazines under the detonating effects of the ZZ rays. [22]

In the same way that ZZ rays can replace conventional ship-to-ship guns, so anti-aircraft guns are replaced by Z rays:

> To guard against spying from aircraft, Z rays—a mild form of the terrible radio-electric current—were directed skywards. Should an aeroplane incautiously or daringly pass within the limits of the Z rays the sudden failure of the motor would result, and the aircraft would be compelled to come to earth. [23]

The real action starts when a mad scientist named Georgeos Kosmosoli develops even more powerful ray weapons of his own: "Compared with these

Fig. 5 The cover of *Amazing Stories* for August 1927, illustrating a scene from H. G. Wells's 1897 novel *The War of the Worlds*—complete with Martian "heat ray" weapons (public domain image)

rays the British ZZ rays are but a candle to an arc lamp" [24]. Their effect is to disintegrate any metal object that comes within range. Kosmosoli also has a counterpart to the Z rays, which shuts down not just aircraft engines but any electrical systems within its sphere of influence. This allows him to hold the world to ransom:

> The lesser rays were bad enough. Modern civilization had grown so dependent upon the telephone, telegraphy, internal combustion engines electrically fired, electric light, and a hundred other inventions of a kindred nature, that the sudden and total paralysis of the electric world ... meant a great financial loss. [25]

Needless to say, British ingenuity eventually wins the day—by returning to the weapons and tactics of the Middle Ages—and Kosmosoli is defeated.

By the 1930s, ray-based weapons had become inextricably associated with the pulp science fiction magazines pioneered by Hugo Gernsback—to the extent that author William Gibson, in his short story "The Gernsback Continuum" (1981), coined the term "ray gun gothic" to describe that whole sub-genre of SF [26].

Even Gernsback himself was conscious that ray guns were becoming an SF cliché. Introducing John W. Campbell's story "Space Rays" in the December 1932 issue of *Wonder Stories*, he wrote:

> Campbell ... has proceeded in an earnest way to burlesque some of our rash authors to whom plausibility and possible science mean nothing. He pulls, magician-like, all sorts of impossible rays from his silk hat, much as a magician extracts rabbits. [27]

Whether Campbell really meant his story to be taken as a parody is debatable, but that's certainly the way it comes across in hindsight. The story's villain, a space pirate, has armed himself with a whole arsenal of colour-coded rays, from a blue-green paralyzer ray and an orange pain-stimulating ray to a top-of-the-range lethal white ray.

As archetypally sci-fi as the idea of a "death ray" is, there was a time—in the years before World War Two—when it wasn't confined to works of fiction. It was also the subject of more serious speculation, and even a number of practical tests that were supposed to have given promising results. One of the first people to make such claims was the inventor Nikola Tesla, who popped up a couple of times in the previous chapter, "Electronic Brains". As early as 1915 he wrote:

> It is perfectly practical to transmit electrical energy without wires and produce destructive effects at a distance. I have already constructed a wireless transmitter which makes this possible. Where unavoidable, the transmitter may be used to destroy property and life. [28]

Within a few years, other people were making similar claims. On 28 May 1924, for example, the *New York Times*, referring to the Soviet politician Leon Trotsky, reported as follows:

> News has leaked out from the communist circles in Moscow that behind Trotsky's recent war-like utterances lies an electromagnetic invention, by a Russian engineer named Grammachikoff for destroying airplanes. [29]

Britain, too, had its counterpart to Tesla and Grammachikoff, in the person of Harry Grindell Matthews—or "Death Ray Matthews", as he was dubbed by the media of the time. In an article of that title in the September 2003 issue of *Fortean Times*, David Clarke and Andy Roberts explain how, in the autumn of 1923, "select journalists were given a demonstration of Matthews's ray stopping a motor cycle engine at a distance of 15 metres" [30].

That sounds a lot like the fictional "Z rays" in Percy F. Westerman's *War of the Wireless Waves*—which was (by coincidence or otherwise) published in the very same year. But what of Westerman's more powerful "ZZ rays", which could remotely detonate an enemy ship's munitions store? Here is Clarke and Roberts's account of another demonstration by Grindell Matthews, in April 1924:

A wide-eyed *London Star* reporter was ushered into Matthews's London laboratory and shown a bowl of gunpowder being ignited by the ray. Matthews was at pains to explain this was only the beginning, a small scale demonstration of what could easily be the destruction of ammunition dumps at huge distances or the destruction of aeroplane engines in flight. [30]

The British media loved Grindell Matthews, but officialdom was less impressed. In May 1924, a government minister told the House of Commons:

We are not in a position to pass judgment on the value of this ray, because we have not been allowed to make proper tests. Therefore whether there is anything in it or not still remains unexplored. The departments have been placed in a difficult position in dealing with the matter partly because of the vigorous press campaign conducted on behalf of this gentleman, and partly because this is not the first occasion on which the inventor has put forward a scheme for which extravagant claims have been made. The result is the departments are not able to accept Mr Grindell Matthews's statement about this invention without a scrutiny which he is not prepared to face. [30]

As Clarke and Roberts conclude:

Matthews never managed to successfully demonstrate his invention to anyone's satisfaction. Whether this was because it was a complex money-making scam or whether the world's governments were incapable of grasping the enormity of his ideas is unclear. [30]

Nevertheless, the affair did lead the British government to take one history-making step, as the two authors go on to point out:

On 18 January 1935 H. E. Wimperis, the director of scientific research at the Air Ministry, approached scientist Robert Watson-Watt to advise the government "on the practicability of proposals of the type colloquially called death ray." [30]

The significance of this lies not in the government's request to Watson-Watt, but in his response to it. He realized immediately that what they were asking was impractical, for reasons of simple geometry. Radio energy spreads out in all directions, which is why it's possible to broadcast a signal to a large area. For the same reason, however, a radio transmission would never have the concentrated power needed to do appreciable damage, such as downing an aircraft in flight. Watson-Watt suggested the government's money would be better spent in working out how to detect an aircraft—rather than destroy it—by directing radio energy at it.

That's the idea behind radar, of course, and it wasn't new. As we saw in the previous chapter, the same suggestion had been made by Tesla as far back as 1900. The difference this time was that the government took the idea seriously, and gave Watson-Watt the necessary funds to make it work. The first military radar system, called Chain Home, was in place along the east coast of Britain by 1939, just in time for World War Two.

If radio waves won't work as a death ray, what will? Science fiction was traditionally rather vague about the physics of ray guns—which was, in essence, Hugo Gernsback's criticism of John W. Campbell's "Space Rays" story in 1932. A decade later, however—when Campbell was the editor of his own magazine, *Astounding Science Fiction*—he redeemed himself by printing a story that contained a detailed description of a beam weapon that really would have worked.

To backtrack a bit—during a large part of the 20th century, there was a very common piece of technology with the sci-fi sounding name of "electron gun". Most people had one, but they were scarcely aware of it because it was sealed inside the glass of their TV tubes—or any other "cathode ray tube".

As mentioned earlier, cathode rays consist of a tight beam of electrons inside a vacuum tube. It's this beam that the electron gun produces, and it has to be confined inside a tube because it wouldn't work in the Earth's atmosphere. There's no atmosphere in outer space, however—and so no necessity for a confining tube. In space, in other words, an electron gun could be used as a real weapon. That's the idea behind George O. Smith's story "Recoil", in the November 1943 *Astounding* (see Fig. 6).

The story is one of a series Smith wrote set on a space station called Venus Equilateral. One of the hazards it faces is the occasional meteor impact, and engineer Walt Franks sets out to design a weapon that can destroy any space

Fig. 6 The cover of *Astounding Science Fiction* for November 1943, featuring the realistically described beam weapon from George O. Smith's short story "Recoil" (public domain image)

debris before it hits. Another character—meta-fictionally aware that the situation is verging on a sci-fi cliché—remarks that: "Walt has turned Buck Rogers and is now about to invent a ray gun." Nevertheless, the idea is perfectly feasible, as Franks explains:

> I've designed an electron gun. It is a superpowered, oversized edition of the kind they used to use in kinescope tubes, oscilloscope tubes, and electron microscopes. Since the dingbat is to be used in space, we can leave the works of the gun open and project a healthy stream of electrons at the offending object without their being slowed and dispersed by an impeding atmosphere. [31]

"Kinescope" is an old word for a TV tube—which Franks's space gun closely resembles, except for the lack of a glass tube and its enormous size:

The cathode is a big affair six feet in diameter and capable of emitting a veritable storm of electrons. We accelerate them by means of properly spaced anodes of the proper voltage level, and we focus them into a nice bundle by means of electrostatic lenses. [31]

The last point is critical to the functioning of any electron gun, real or fictitious. The electrons have to be kept in a tight beam—a process called collimation—in order to avoid the same sort of spreading-out that dissipates energy in the case of a radio transmission.

There's another important factor which makes Smith's fictional electron gun a viable proposition. The space station happens to be located a long way from any planet or other source of magnetic fields; it orbits the Sun at the same distance as Venus, but 60 degrees ahead of it (the name "Venus Equilateral" comes from the fact that the station, Sun and Venus form the corners of an equilateral triangle). The lack of any magnetic field means the electron beam can travel in a dead straight line.

That wouldn't be the case for a similar gun located in Earth orbit—where it would have to be, of course, if it were to be of any use against ballistic missiles. As with any electrically charged particle, the path of an electron is deflected by a magnetic field. That's the very principle that made it possible to form an image on an old-style cathode ray tube. It's also the reason the idea of a charged particle beam was deemed impractical as a space weapon—even though it was given serious consideration. In the words of Roald Sagdeev, a former space scientist at the USSR Academy of Sciences:

The original concept was that you would accelerate charged particles to high energy. The main problem was that you would be unable to fire it against a target because it would be deflected by the Earth's natural magnetic field. [32]

The same problem doesn't arise in the case of electromagnetic radiation, because the particles involved—photons—are electrically neutral. As mentioned earlier, however, they suffer from a different but equally serious problem—the way a beam spreads out with distance, preventing it from delivering a concentrated amount of energy onto a single target. The ideal solution would be if an electromagnetic beam could be "collimated" in the same way that an electron beam is. As impossible as that seemed in the first half of the 20th century, it finally became reality with the invention of the laser.

As it happens, the essential secret of lasers—a process called stimulated emission—was discovered by Albert Einstein as long ago as 1916. In simple terms, this involves a photon interacting with an atom in a way that produces a second photon with exactly the same frequency, polarization and—most importantly—direction of travel as the first photon. For various reasons, however, many years passed before anyone could see how to turn Einstein's theory into a practical reality.

The breakthrough was made in 1957 by Gordon Gould—a scientist working for the American defence contractor TRG. As science writer Brian Clegg explains:

> Gould realized there was a potential for pass after pass . . . to build up more and more stimulation, eventually producing an intensely concentrated beam of light. He believed it should be possible to create a beam that would easily produce temperatures as high as the Sun's surface—around 5,500°C. . . . Gould headed his notes: "Some rough calculations on the feasibility of a LASER: Light Amplification by the Stimulated Emission of Radiation." [33]

Gould quickly convinced himself of the laser's potential power. It could, for example, "punch holes in metal with the sheer concentrated energy of its beam of light" [34]. It was the closest the real world had ever come to a truly practical death ray.

The military relevance of Gould's work was obvious—and very timely, since it coincided with the Sputnik crisis, and the prospect of a war in space that would be fought with radically new technologies. To quote Brian Clegg again:

> Until this point in history it had been difficult to interest the conservative US military in leading-edge technology, but the proposal arrived at an opportune moment. In response to the shock of the Soviet Sputnik becoming the first man-made satellite, flaunting apparent Soviet technological superiority, the US government had set up the Advanced Research Projects Agency. ARPA was to fund and manage precisely the kind of crazy possibilities that Gould was claiming for his laser. Gould responded by dreaming up other potential military uses for his device . . . the power of lasers could even be used to destroy incoming missiles. This possibility led to the Air Force giving the project the codename Defender. [35]

As promising as the new laser technology was, it still had to conform to the laws of physics. As with any real-world device, a laser can't throw out more energy than it consumes. That means that any laser powerful enough to be used as a weapon is going to be big—certainly much bigger than the traditional hand-held ray gun of science fiction.

As for SF itself—it had never been bothered by nit-picking considerations like the conservation of energy, and it wasn't going to start now. There was no

Fig. 7 A publicity shot of Captain Kirk holding a phaser rifle, from an early episode of *Star Trek* (public domain image)

let-up in the depiction of ray-based small arms, culminating in the iconic phasers of *Star Trek* (see Fig. 7).

The name "phaser", in fact, was a last-minute afterthought; the original intention had been to call the weapons "lasers". *Star Trek* creator Gene Rodenberry explains how the change came about:

> We were two days into filming on the second pilot when we realized that lasers might very well become commonplace by the time the show got on the air, or at least within the next couple of years. Rather than run the risk of being outdated, we decided to say "phaser" instead. . ... We didn't want people saying to us three years from now, "Oh, come on now, lasers can't do that." [36]

A Shield in Space

As mentioned at the end of the first chapter, "The Super-Bomb", the only really effective defence against nuclear missiles would be some kind of force field—and unfortunately that's just sci-fi technobabble. In the absence of anything remotely like a force field in the real world, huge efforts have been made to find some other technology that doesn't have the same weakness as ABMs—namely the way they can be defeated simply by throwing more warheads at them. Reference has already been made to ARPA's "Project Defender"—which looked at the use of lasers for ballistic missile defence in the early 1960s—but that was just the tip of the iceberg.

Other ARPA projects of a similar vintage include the sci-fi sounding Space Patrol Active Defence (SPAD)—which would have intercepted missiles in their exoatmospheric phase—and the much less excitingly named BAMBI, for Ballistic Missile Boost Intercept. On the subject of the latter, Herb York—a director of ARPA's successor DARPA (Defence Advanced Research Projects Agency)—was pretty scathing:

> The concept of this mad scientist's dream involved surrounding the Earth with a great swarm of small satellites that would detect, attack and destroy anything that stuck its nose above the atmosphere. Some thought had been given to the question of how to enable friendly missiles and satellites to pass through the swarm, but that was one of the least of the problems with this system, and it never got beyond the study stage. [37]

By the 1970s, the number of options still being considered by ARPA had shrunk to just two—high-energy lasers and particle beams—"which most scientists regarded as science fiction", in the words of historian Audra J. Wolfe [38].

One of the strongest advocates of space-based missile defence throughout the 1970s was Edward Teller, who featured prominently in the first chapter, "The Super-Bomb". Teller and a team working at Lawrence Livermore National Laboratory, near San Francisco, came up with Project Excalibur: a super-powerful laser that operated at X-ray frequencies. Triggering it required an unprecedented amount of energy—but with his Manhattan Project heritage, Teller knew the perfect power source for it. As a 1984 report produced for the US government explains:

> The pumping source for the X-ray laser is a nuclear bomb. The radiant heat of the bomb raises electrons to upper energy levels in atoms of lasant material

positioned near the detonation. . . . Since X-rays are not back-reflected by any kind of mirror, there is no way to direct the X-rays into a beam with optics like the visible and infrared lasers. Nonetheless, some direction can be given to the laser energy by forming the lasant material into a long rod. [39]

On paper, this is actually quite a good idea. It's designed to attack ICBMs in the boost phase, before they have a chance to deploy multiple warheads. If a single bomb is used to power a cluster of laser rods, each of those rods can be aimed at a different missile. That makes it much harder to overwhelm the system with sheer numbers, in the way that ABMs or less powerful conventional lasers can be. As the same report says:

Simultaneous launch of all Soviet boosters is not a problem for X-ray lasers in the way it is a problem for chemical lasers that must dwell on each target before passing on to the next. [39]

As promising as it might have been from a technical point of view—and it's fair to say not everyone was convinced of that point—Teller's system was highly unlikely to have made it off the ground for a much more basic reason. What was that clause in the 1967 Outer Space Treaty again? "Parties to the Treaty undertake not to place in orbit around the Earth any objects carrying nuclear weapons" [14].

Whatever its pros and cons, no one can deny that Project Excalibur made for some pretty impressive—and very sci-fi looking—"artist's impressions" (see Fig. 8).

Teller's work on Project Excalibur might have been forgotten altogether, if it hadn't attracted the attention of a local politician—who went on to become the most influential ally Teller could have hoped for. As science writer Nigel Hey recounts:

In 1977, Ronald Reagan, the newly elected governor of California and thus a regent of the University of California, visited the UC-managed Lawrence Livermore National Laboratory at the invitation of its founder, Edward Teller. There, in Teller's words, "I gave him a proposal of a primitive strategic defence initiative. Governor Reagan was very interested. In the end he gave strong support to missile defence." [40]

The relationship continued after Reagan became the 40th President of the United States in 1981. To quote the *New Yorker*, reporting on a meeting that took place in September 1982:

Fig. 8   An artist's impression of Project Excalibur, a space-based X-ray laser designed to neutralize even the most massive ICBM attack (public domain image)

> Dr Teller represented all that the President admired in a scientist, being distinguished, individualistic, sonorously spoken and short of academic circumspection. For half an hour, Teller deployed X-ray lasers all over the Oval Office, reducing hundreds of incoming Soviet missiles to radioactive chaff. [41]

Teller succeeded in persuading Reagan that his X-ray lasers could provide America with an impenetrable "shield in space" against enemy missiles. The result was the most famous televised address of his presidency, on 23 March 1983:

> Let me share with you a vision of the future which offers hope. It is that we embark on a programme to counter the awesome Soviet missile threat with measures that are defensive. Let us turn to the very strengths in technology that spawned our great industrial base and that have given us the quality of life we enjoy today. What if free people could live secure in the knowledge that their security did not rest upon the threat of instant US retaliation to deter a Soviet attack, that we could intercept and destroy strategic ballistic missiles before they reached our own soil or that of our allies? [42]

The next morning, the *New York Post* carried a banner headline yelling "STAR WARS PLAN TO ZAP RED NUKES" [43]. It was a nickname that took hold immediately. To the people involved in it, the programme Reagan had instigated was the Strategic Defence Initiative (SDI). To everyone else, it was "Star Wars"—a name taken from America's most famous sci-fi movie franchise. On closer inspection, however, SDI's links with the world of science fiction go considerably deeper than its popular nickname.

"Would President Reagan ever have made his famous *Star Wars* speech of 23 March 1983 if he hadn't seen so many movies?", Arthur C. Clarke once asked [44]. It's not an inappropriate question. Before Reagan went into politics in the 1960s, he'd been a Hollywood actor. Most of his appearances were in westerns and war films, and none of them in sci-fi—but he did have an interest in the latter. This became clear, as David Clarke explains, in a meeting between Reagan and his Soviet counterpart Mikhail Gorbachev:

> According to Lou Cannon's biography of Reagan, the 40th President of the United States was an avid science fiction fan during his Hollywood years. His favourite genre was "the invasion from outer space that prompts Earthlings to put aside nationalistic quarrels and band together against the alien invader". Reagan liked this idea so much that in 1985 he surprised Mikhail Gorbachev at a summit in Geneva by saying he was confident the two superpowers would cooperate if Earth were threatened by alien invasion. Taken aback, the Soviet leader changed the subject. [45]

Whether Reagan's "Star Wars" speech was inspired by SF or not, the genre really did have a foot in the SDI door—in the person of Jerry Pournelle and Larry Niven, who co-authored a number of bestselling novels including *The Mote in God's Eye* (1974) and *Lucifer's Hammer* (1977). As Nigel Hey recounts:

> Some of the rights to claim authorship of SDI belong to Jerry Pournelle ... Pournelle's Citizen's Advisory Council on National Space Policy, formed in 1980, generally met at the Tarzana, California home of his colleague and co-author, Larry Niven. These were large and diverse get-togethers catered lavishly by Niven's wife Marilyn, with attendees from the military, science fiction, aerospace and science communities. There, astronauts Buzz Aldrin and Pete Conrad were mixing with the likes of ... Greg Bear and Robert Heinlein. [46]

Hey goes on to quote Pournelle himself: "I think you will find that most of the intellectual firepower behind SDI was in meetings in Larry Niven's house in Tarzana between 1980 and 1984" [46].

Even before then, space-based defences that looked uncannily like SDI had made an appearance in fiction. There's Ben Bova's 1976 novel *Millennium*, for example. Set, as the title suggests, in 1999, it shows America in the process of setting up a "globe-spanning network of laser-armed satellites" [47]. At the same time, a still-extant Soviet Union is busy destroying them—as the story's Defence Secretary explains to the President:

> We have been trying to complete deployment of our strategic defence network for the past two years. The Soviets have been incapacitating our satellites to prevent us from finishing the system. . . .. You will see that they are now knocking out our satellites almost as fast as we can launch them. [48]

Another work of the 1970s that depicted a similar situation was General Sir John Hackett's novel *The Third World War* (1979). As mentioned in the second chapter, "Journey into Space", this describes a fictional East-West war in the then-future of 1985. Outlining the events leading up to war, Hackett again presents a pre-SDI vision of space-based lasers:

> In the late seventies . . . the USSR continued with frequent manned launches on research tasks, the exact purposes of which were not always evident to Western observers. It was however, pretty well authenticated that they were developing a counter-satellite capability employing, in addition to jamming, certainly lasers and other high-energy beams. In 1985 both sides had some counter-satellite capability, but it was suspected that the USSR was well in the lead. [49]

The assertion that "the USSR was well in the lead" may look wrong in a post-Reagan light—but bear in mind that Hackett was writing before Reagan had even been elected. In the 1970s, the Russians were indeed more proactive than the Americans when it came to developing space weapons, as this quote from Nigel Hey makes clear:

> In 1976, the space side of the Soviet defence industry had been secretly tasked with the job of developing a large manned orbital battle station, which if needed (and if the technology were available) could carry small laser weapons, anti-satellite cannons, small rockets, even nuclear weapons. Given the code name Skif, it would be a much scaled up descendant of the earlier Almaz. [50]

As mentioned earlier in this chapter, Almaz was the military variant of the Salyut space station. Its larger successor, Skif, never materialized. A preliminary test version, called Polyus, was launched in May 1987, but it failed to reach orbit. As space historian Paul Drye relates:

> Polyus carried a wide variety of experiments, but its main purpose was to test the Skif-DM, a 1-megawatt carbon dioxide gas laser weapon that the USSR had been working on since 1983. . . .. Various components of the laser were deliberately left

Fig. 9 A US Defence Intelligence Agency illustration of a proposed Soviet ground-based laser, which would have been used to destroy satellites in orbit (public domain image)

out, leaving it with only the ability to track targets, and Gorbachev is reported to have banned testing of even that during a visit to Baikonur a few days before launch. [51]

Although it was designed at the same time as Teller's super-powerful X-ray laser, the Soviet laser referred to here used a completely different approach, working at much lower, infrared, frequencies. It was nothing like as powerful as the X-ray laser, but had the advantage that it could—if its development had gone ahead—have been used as a ground-based anti-satellite weapon as well as a space-based one (see Fig. 9).

As much as they've been studied and talked about, no anti-missile laser—or anti-satellite laser—has ever been deployed operationally. That's essentially been the story of this whole chapter, from orbiting bombs and armed Soyuz fighters to air-launched ASAT missiles and Ronald Reagan's grand vision of SDI. All those things ended up being cancelled. They have another thing in common, too—they all sound like science fiction. Yet they were real . . . or were they?

Looking back on the SDI project, one of the scientists involved in it, Gerold Yonas, made the following intriguing remarks:

I don't know any technical person who believed all of the stuff we said. We knew it was a game. [52]

And:

We were lying to the Russians. They were lying to us. The Cold War was characterized by deception on all levels. [53]

This brings us on to a whole new aspect of the Cold War—the prominent role played by disinformation and other mind games. That's the subject of the next chapter.

References

1. T. Burnett, *Who Really Won the Space Race?* (Collins & Brown, London, 2005), pp. 204, 205
2. T. Burnett, *Who Really Won the Space Race?* (Collins & Brown, London, 2005), p. 169
3. A. MacLean, *Ice Station Zebra* (Fontana, London, 1965), p. 233
4. D.A. Day, Has anybody seen our satellite? *The Space Review* (April 2009), http://www.thespacereview.com/article/1352/1
5. D.A. Wollheim, *Mike Mars and the Mystery Satellite* (Paperback Library, New York, 1966), p. 101
6. D.A. Wollheim, *Mike Mars and the Mystery Satellite* (Paperback Library, New York, 1966), p. 103
7. N. Hey, *The Star Wars Enigma* (Potomac Books, Washington DC, 2007), p. 30
8. N. Hey, *The Star Wars Enigma* (Potomac Books, Washington DC, 2007), p. 39
9. K. Neville, Cold War, in *The Astounding Science Fiction Anthology*, ed. by J.W. Campbell (Simon & Schuster, New York, 1952), pp. 404–414
10. J. Sutton, *Bombs in Orbit* (Ace Books, New York, 1959), pp. 14, 15
11. J. van Hise, *The Unauthorized History of Trek* (Harper Collins, London, 1997), p. 48
12. K.W. Gatland, *Development of the Guided Missile* (Iliffe & Sons, London, 1954), pp. 211, 212
13. GR-1/SS-X-10 SCRAG, Federation of American Scientists, https://fas.org/nuke/guide/russia/icbm/gr-1.htm
14. Outer Space Treaty of 1967, Wikisource, https://en.wikisource.org/wiki/Outer_Space_Treaty_of_1967
15. D. Baker, *Soyuz 1967 Onwards* (Haynes, Yeovil, 2014), pp. 103–105
16. P. Drye, *False Steps: The Space Race as it Might Have Been* (Baggage Books, Kindle edition, 2015), loc. 2483
17. K.P. Werrell, *Hitting a Bullet with a Bullet: A History of Ballistic Missile Defence* (Air Power Research Institute, 2000), http://www.dtic.mil/get-tr-doc/pdf?AD=ADA381863

18. J.P. Martino, Pushbutton War, in *Prologue to Analog*, ed. by J.W. Campbell (Panther, London, 1969), pp. 198–215
19. N. Hey, *The Star Wars Enigma* (Potomac Books, Washington DC, 2007), p. 34
20. I. Asimov, *Biographical Encyclopedia of Science and Technology* (Pan Books, London, 1975), p. 402
21. I. Asimov, *Biographical Encyclopedia of Science and Technology* (Pan Books, London, 1975), p. 442
22. P.F. Westerman, *The War of the Wireless Waves* (Oxford University Press, London, 1936), p. 13
23. P.F. Westerman, *The War of the Wireless Waves* (Oxford University Press, London, 1936), pp. 14, 15
24. P.F. Westerman, *The War of the Wireless Waves* (Oxford University Press, London, 1936), pp. 92, 93
25. P.F. Westerman, *The War of the Wireless Waves* (Oxford University Press, London, 1936), p. 215
26. W. Gibson, The Gernsback Continuum, in *Burning Chrome* (Grafton, London, 1988) pp. 37–50
27. J.W. Campbell, Space Rays, in *Wonder Stories* (December 1932), pp. 585–593
28. G. Owen, Directed-energy weapons: a historical perspective. J. Def. Sci. **2**(1), 89–93 (1997)
29. N. Redfern, *Science Fiction Secrets* (Anomalist Books, San Antonio, 2009), p. 112
30. D. Clarke, A. Roberts, Death Ray Matthews, in *Fortean Times* (September 2003), pp. 38–47
31. G.O. Smith, *Venus Equilateral*, vol 1 (Orbit Books, London, 1975)
32. N. Hey, *The Star Wars Enigma* (Potomac Books, Washington DC, 2007), p. 23
33. B. Clegg, *The Quantum Age* (Icon Books, Kindle edition, 2014), loc. 1586–1592
34. B. Clegg, *The Quantum Age* (Icon Books, Kindle edition, 2014), loc. 1602
35. B. Clegg, *The Quantum Age* (Icon Books, Kindle edition, 2014), loc. 1610–1614
36. S.E. Whitfield, G. Roddenberry, *The Making of Star Trek* (Ballantine, New York, 1968), p. 166
37. N. Hey, *The Star Wars Enigma* (Potomac Books, Washington DC, 2007), pp. 24–26
38. A.J. Wolfe, *Competing with the Soviets* (Johns Hopkins University Press, Baltimore, 2013), p. 129
39. A.B. Carter, *Directed Energy Missile Defense in Space* (U.S. Congress Office of Technology Assessment, 1984), http://www.princeton.edu/~ota/disk3/1984/8410/8410.PDF
40. N. Hey, *The Star Wars Enigma* (Potomac Books, Washington DC, 2007), p. 53
41. N. Hey, *The Star Wars Enigma* (Potomac Books, Washington DC, 2007), p. 81
42. N. Hey, *The Star Wars Enigma* (Potomac Books, Washington DC, 2007), pp. 93, 94
43. N. Hey, *The Star Wars Enigma* (Potomac Books, Washington DC, 2007), p. 95
44. A.C. Clarke, *Astounding Days* (Gollancz, London, 1990). pp. 97, 98

45. D. Clarke, *How UFOs Conquered the World* (Aurum Press, London, 2015), p. 14
46. N. Hey, *The Star Wars Enigma* (Potomac Books, Washington DC, 2007), p. 73
47. B. Bova, *Millennium* (Methuen, London, 1988), p. 46
48. B. Bova, *Millennium* (Methuen, London, 1988), pp. 13–14
49. G.S.J. Hackett, *The Third World War* (Sphere, London, 1979), p. 255
50. N. Hey, *The Star Wars Enigma* (Potomac Books, Washington DC, 2007), p. 51
51. P. Drye, *False Steps: The Space Race as it Might Have Been* (Baggage Books, Kindle edition, 2015), loc. 3079
52. N. Hey, *The Star Wars Enigma* (Potomac Books, Washington DC, 2007), p. 210
53. N. Hey, *The Star Wars Enigma* (Potomac Books, Washington DC, 2007), p. 223

Mind Games

In which the covert fringes of the Cold War see a blurring of fact and fiction, as intelligence agencies experiment with sci-fi-sounding hypnosis machines and mind-control drugs—the success or failure of which is largely a matter of opinion. The same period also saw science fiction itself exploited for propaganda purposes, while popular SF tropes like antigravity provided suitably opposition-confusing material for the Cold War disinformation specialists.

High-Tech Hypnosis

The use of hypnotism to make people do things against their will has always been a common ploy among the villains of popular fiction. An archetypal proponent of such tactics was Sax Rohmer's Fu Manchu, who generally achieved the desired hypnotic state with the aid of drugs and other chemicals. As he explains to a prospective victim in the 1932 movie *The Hand of Fu Manchu*:

> This serum, distilled from dragon's blood, my own blood, the organs of different reptiles, and mixed with the magic brew of the sacred seven herbs, will temporarily change you into the living instrument of my will. You will do as I command! [1]

The science fiction genre, on the other hand, with its predilection for high-tech hardware, tended to spurn hypnotic drugs in favour of "hypnotic ray"

© Springer International Publishing AG, part of Springer Nature 2018
A. May, *Rockets and Ray Guns: The Sci-Fi Science of the Cold War*, Science and Fiction,
https://doi.org/10.1007/978-3-319-89830-8_5

Fig. 1 Two panels from the February 1942 issue of *Prize Comics*, showing the villainous Dr Mesmeric using his hypnotic ray machine on the hero, Power Nelson (public domain image)

machines. By the middle of the 20th century, these had become a common trope in the lower strata of pulp fiction and comics (see Fig. 1).

The concept of a "hypno-ray" is still alive and well today—though usually in a tongue-in-cheek context, as the *TV Tropes* website explains:

> A standard gadget in the Mad Scientist grab bag, the hypno-ray allows the user to instantly hypnotize an unsuspecting target and make that person do their evil bidding. . ... These days, this type of device is usually limited to comedic settings. [2]

In the real world, talk about mind-control rays is most often found among people who believe themselves to be on the receiving end of them—and who may go to the extent of covering their head with aluminium foil in order to protect themselves. Both these behaviours—complaining about mind-control rays and wearing "tinfoil hats"—are recognized symptoms of paranoia.

From a science-fictional point of view, one person who suffered from such symptoms was particularly significant. This was a factory worker and would-be artist named Richard Shaver, who started to believe he was under malevolent mind control in the early 1930s. He was declared "mentally incompetent" and institutionalized in a Michigan state hospital in 1934 [3].

After his release, Shaver began a correspondence with the editor of *Amazing Stories*, Ray Palmer, in 1943. He sent him long rambling accounts of supposedly real experiences, which Palmer edited and recast in the form of SF adventure stories. The first of these was a novella called "I Remember Lemuria", which Palmer printed in the March 1945 issue of *Amazing*. It told of a network of subterranean caves populated by a race of evil, degenerate humanoids called "deros", who used high-tech ray machines to control the surface population.

As far as Shaver was concerned, all this was factually true. As he asserts in a footnote to the story:

> The course of history, the battles, the decisions of tyrants and kings—was almost invariably decided by interfering control from the caverns and their hidden apparatus. . . .. The remarkable part of it all is that it still goes on today. Emotional and mental stim—unsuspected by such as you and the average citizen—used in mad prankishness, all come from the ancient apparatus. . . .. The dero of the caves are the greatest menace to our happiness and progress; the cause of many mad things that happen to us, even so far as murder. Many people know something of it, but they say they do not. They are lying. They fear to be called mad, or to be held up to ridicule. Examine your own memory carefully. You will find many evidences of outside stim. [4]

As Palmer's biographer Fred Nadis pointed out, "many of Shaver's assertions indicate classic symptoms of paranoia" [5]. In spite of that—or perhaps because of it—"I Remember Lemuria" proved to be one of the most successful stories *Amazing* had ever published. Not everyone liked it, but everyone had an opinion about it, and it boosted circulation enormously.

Latching onto a good thing, Palmer went on to print another 20-plus "Shaver Mystery" stories in the magazine. The new readers drawn in by these tended to view them as "fact" rather than fiction, and eagerly sent in mind-control anecdotes of their own. By August 1946, Palmer claimed the magazine had received "over 10,000 confirming letters from our readers" [6].

Traditional SF fans, on the other hand, were more likely to hate the Shaver stories. A prominent fanzine of the time complained that they were turning SF "into a plaything for every semi-sane crackpot" [7]. Another leading fan, Forrest J. Ackerman, circulated a petition demanding that *Amazing* put an end to the Shaver Mystery. This sustained pressure had its effect on *Amazing*'s publishers, and by the end of the 1940s the magazine was no longer publishing Shaver—and Palmer was no longer its editor.

While the Shaver Mystery fell into oblivion, the idea of mechanically assisted mind control didn't. For modern conspiracy theorists, the sinister power behind such activities isn't a bizarre subterranean race, it's the government. As paranoid as this idea sounds, it's not entirely unjustified—at least, not in the context of that most paranoid of all times, the Cold War.

From quite early in its existence, the CIA really did explore the possibilities of mind control—as a weapon to use against the enemy. They started off with old-fashioned hypnosis, funding a study in 1956, "to investigate the possibilities of hypnotic induction of non-willing subjects" [8].

It wasn't long before the idea of using hypnosis for military purposes entered mainstream consciousness, in the form of *The Manchurian Candidate*—both the 1959 novel by Richard Condon and the 1962 film starring Frank Sinatra. The plot centres around the use of post-hypnotic commands to turn one of characters—an American former prisoner-of-war—into a "programmed assassin".

Apart from its Cold War setting, *The Manchurian Candidate* is old hat—at least as far as fiction is concerned. A very similar scenario appears in Sax Rohmer's book *The Si-Fan Mysteries* (1917), which features the narrator Dr Petrie being drugged and hypnotized by an ally of Fu Manchu named Ki-Ming. Soon after, Petrie attempts to shoot his friend Nayland Smith—Fu Manchu's arch-nemesis—under the delusion that Smith is Fu Manchu himself. The ploy doesn't work, because Smith took the precaution of disabling Petrie's gun. After Petrie has recovered, the situation is explained to him by a medical specialist:

> You see, Dr. Petrie, you were taken into the presence of a very accomplished hypnotist, having been previously prepared by a stiff administration of *maagûn*. You are doubtless familiar with the remarkable experiments in psycho-therapeutics conducted at the Salpêtrier in Paris, and you will readily understand me when I say that, prior to your recovering consciousness in the presence of the mandarin Ki-Ming, you had received your hypnotic instructions. These were to be put into execution either at a certain time (duly impressed upon your drugged mind) or at a given signal. [9]

If the story had been science fiction, the drug *maagûn* would have been replaced with a hypno-ray. Although many readers would find that more exciting, it's actually a lot less credible. You really can affect the mind with drugs, but it can't be done with a ray machine. Or can it? Surprisingly, a series of experiments during the Cold War came close to doing just that.

In the 1960s, America's Office of Naval Intelligence funded a research project by the neuroscientist Jose Delgado, who believed that "motion, emotion and behaviour can be directed by electrical forces, and humans can be controlled like robots by push buttons" [10]. Describing some of his experiments in a scientific paper in 1968, Delgado wrote:

Radio stimulation on different points in the amygdala and hippocampus in the four patients produced a variety of effects, including pleasant sensations, elation, deep thoughtful concentration, odd feelings, super relaxation, coloured visions and other responses. [10]

Unfortunately, as far as a practical hypno-ray was concerned, there was a catch. Although Delgado obtained his results using a radio transmitter, the subjects first had to be implanted with an electronic device called a "stimoceiver"—a stimulus receiver. Once that was done, the effect could be dramatic in the extreme—as in a video Delgado produced showing him stopping a charging bull in its tracks [10]. The bull, however, had to be fitted with a stimoceiver, or the trick wouldn't have worked.

Technically, this is mind control—but only in the limited sense of controlling a mind that has previously been implanted with a suitable electronic receiver. That's a far cry from the indiscriminate "broadcast" mind control that conspiracy theorists talk about, which works on anyone who happens to be within range.

Some people believe that this, too, went on during the Cold War. For more than two decades, from 1953 to 1976, the US Embassy in Moscow was deliberately bathed in microwave radiation by the Soviet authorities. No one knows for certain why this was, but it allegedly had adverse effects on some of the embassy staff.

At least some people in the US government were worried the microwaves represented some type of psychic weapon. As a result, Project Pandora was set up by DARPA—the Defence Advanced Research Projects Agency—to, in the words of a declassified memo, "discover whether a carefully controlled microwave signal could control the mind" [10].

This much is fact—but it may not mean anything. Contrary to what many conspiracy theorists believe, the existence of an official investigation into a phenomenon doesn't constitute proof that said phenomenon is real (cf. UFOs, in the next chapter, "Weird Science"). To date, there isn't a shred of evidence in the public domain that microwaves really can affect the human brain—or at least, not in a militarily useful way.

In 1978, an exhaustive study of more than 2000 employees from the Moscow embassy concluded that "there is no convincing evidence to implicate the exposure of these people to microwave radiation and the onset of any adverse health effects" [11]. It's true that some of the test subjects displayed symptoms like anxiety and insomnia—but those symptoms could have had many other causes besides the microwaves. In any case, "anxiety and insomnia" aren't in the same league as a programmed assassin.

So why did the Soviets transmit microwaves at the US embassy? The most likely reason was to jam covert espionage equipment that worked on exactly those frequencies. As journalist Fred Kaplan explains:

> On the 10th floor of the US Embassy in Moscow, the National Security Agency had installed a vast array of electronic intelligence gear. In a city of few skyscrapers, the 10th floor offered a panoramic view, so microwave receivers were able to scoop up phone conversations between top Soviet officials, including Communist Party Chairman Leonid Brezhnev, as they rode around the city in their limousines. . . .. As the Russians realized what the Americans were up to, they returned the gesture, shooting microwave beams at the 10th-floor window. [12]

In this scenario, any adverse health effects caused by the microwaves would merely have been side-effects of their intended use—and not very striking side effects at that. Electronic mind control, of a militarily useful kind, remains the stuff of science fiction. In the real-world, its significance pales in comparison to what really does work, and awesomely so: chemical and biological agents.

Something in the Air

Chemical weapons, in the form of poisonous gases, were used by both sides in World War One. They produced horrifying effects—and that wasn't just true of lethal gases like chlorine, which killed by asphyxiation. Mustard gas, although rarely lethal, produced effects that were like something out of a horror movie. It caused blisters to form all over the skin, painful damage to the lungs—which made victims feel as though they were choking to death, even if they weren't—and blindness.

The one good thing to come out of this was that it turned public opinion firmly against the use of such weapons. As a result, they were prohibited by international law, along with "the use of bacteriological methods of warfare", in the Geneva Protocol of 1925 [13].

It was, however, only the "use" of chemical and biological weapons (CBW) that was forbidden. There was nothing to stop countries researching and stockpiling such weapons, which they continued to do throughout the 20th century. A whole range of Cold War weapon systems were designed to deliver CBW agents, should the Geneva ban ever break down (see Fig. 2).

In the popular imagination, the CBW threat never inspired the same fear as the nuclear one. Nevertheless, biological agents in particular were just as capable of delivering a "doomsday" scenario. The deadliness of a virus or

Fig. 2 A Soviet Mi-24 helicopter deploying a CBW spray, as depicted by a US Defence Intelligence Agency artist in 1986 (public domain image)

other disease-carrying micro-organism can be measured in different ways: the ease with which it multiplies and propagates through the atmosphere, how quickly it spreads from person to person, how likely a victim is to die once infected—and, of course, whether there is any known cure or antidote.

In the worst-case scenario—a rapidly multiplying, highly contagious, extremely deadly virus with no antidote—a biological weapon could depopulate the world as effectively as a nuclear one. That's the idea behind Alistair MacLean's novel *The Satan Bug* (1962), in which a scientist contrasts the effect of such a micro-organism with a more controllable one, botulinus:

> "Botulinus has its drawbacks," Gregori said quietly. "From a military viewpoint, that is. Botulinus you must breathe or swallow to become infected. It is not contagious. . ... Unlike botulinus, this new virus is indestructible—extremes of heat and cold, oxidization and poison have no effect upon it and its life span appears to be indefinite. . ... Unlike botulinus it is highly contagious, as well as being fatal if swallowed or breathed; and, most terrible of all, we have been unable to discover a vaccine for it. . ... To this virus we have given a highly unscientific name, but one that describes it perfectly—the Satan Bug. It is the most terrible and terrifying weapon mankind has ever known or ever will know." [14]

At the other end of the spectrum, and closer to the main topic of this chapter, other CBW agents may offer more subtle military benefits—

including that Holy Grail of "mind control". This was another area the CIA looked at during the Cold War, in a notorious project called MKULTRA. Running from the 1950s to the 1970s, this investigated the potential use of mind-altering drugs in warfare.

A memorandum produced for the director of the CIA in July 1963, ten years after MKULTRA was set up, describes the project as follows:

> The MKULTRA activity is concerned with the research and development of chemical, biological and radiological materials capable of employment in clandestine operations to control human behaviour. The end products of such research are subject to very strict controls including a requirement for the personal approval of the Deputy Director/Plans (DD/P) for any operational use made of these end products. The cryptonym MKULTRA encompasses the research and development phase and a second cryptonym MKDELTA denotes the DD/P system for control of the operational employment of such materials. [15]

MKULTRA's notoriety stems from the fact that much of its testing was done on subjects without their consent. Quoting again from the same memo:

> It is the firm doctrine in Technical Services Division (TSD) that testing of materials under accepted scientific procedure fails to disclose the full pattern of reactions and attributions that may occur in operational situations. TSD initiated a programme for covert testing of materials on unwitting US citizens in 1955. [16]

From the earliest days of MKULTRA, one of the chemical agents favoured by the CIA for "unconventional warfare" was the mind-altering, hallucinogenic drug LSD—later to become just as popular, though for different reasons, with the hippie counterculture of the 1960s. As a CIA memorandum from August 1954 explains:

> Lysergic acid diethylamide (LSD), a drug derived from ergot, is of great strategic significance as a potential agent in unconventional warfare and in interrogations. In effective doses, LSD is not lethal, nor does it have colour, odour or taste. Since the effect of this drug is temporary in contrast to the fatal nerve agents, there are important strategic advantages for its use in certain operations. Possessing both a wide margin of safety and the requisite physiological properties, it is capable of rendering whole groups of people, including military forces, indifferent to their surroundings and situations, interfering with planning and judgment, and even creating apprehension, uncontrollable confusion and terror. [17]

When the existence of the hitherto secret MKULTRA project became public knowledge in the 1970s, the US Congress set up a committee to investigate its activities. Chaired by Senator Frank Church, its report was damning—as the *Washington Post* reported in 1977:

> The Church committee's final report in April 1976 . . . found that MK-ULTRA gave LSD to unwitting subjects (one of whom, Dr Frank Olson, died as a result), used private institutions clandestinely to conduct research, and used prisoners and patients as subjects. [18]

While MKULTRA was chiefly concerned with the targeted administration of drugs to specific individuals—for example prisoners of war—the CIA also considered its indiscriminate use against much larger groups. The idea of the authorities putting "LSD in the water supply" became the stuff of urban legend—but the tactic really was considered for use against enemy forces, as the following excerpt from a CIA document reveals:

> If the concept of contaminating a city's water supply seems, or in actual fact is found to be, far-fetched . . . there is still the possibility of contaminating, say, the water supply of a bomber base or, more easily still, that of a battleship. . . . Our current work contains the strong suggestion that LSD-25 will produce hysteria (unaccountable laughing, anxiety, horror). . . . It requires little imagination to realize what the consequences might be if a battleship's crew were so affected. [19]

The widespread deployment of hallucinogenic chemicals in warfare—by means of air-dropped bombs, rather than the water supply—was the subject of Brian Aldiss's 1969 novel *Barefoot in the Head*. Although it was written in the midst of the Cold War, the novel was unusual for its time in that it didn't visualize its future war in an East-versus-West, communist-versus-capitalist context. Instead—perhaps prophetically—the conflict emerges from the politics of the Middle-East:

> When the Acid Head War had broken out, Kuwait had struck at all the prosperous countries. Britain had been the first nation to suffer the PCA bomb—the Psycho-Chemical Aerosols that propagated psychotomimetic states. [20]

Aldiss doesn't name the specific chemical involved, but "acid" was a common slang term for LSD—and "acid head" for a user of that drug. At another point in the novel, he says:

The PCA bombs had been ideal weapons. The psychedelic drugs concocted by the Arab state were tasteless, odourless, colourless and hence virtually undetectable. They were cheaply made, easily delivered. They were equally effective whether inhaled, drunk or filtered through the pores of the skin. They were enormously potent. The after-effects, dependent on size of dose, could last a lifetime. [21]

The action of *Barefoot in the Head* is set after the end of war, when large parts of Europe are still suffering from its after-effects—producing an anarchistic, spaced-out society in which everyone thinks and talks like a hippie.

A completely different take on a similar theme can be found in James Herbert's 1975 novel *The Fog*. The psychoactive substance in this case is a bacterial agent capable of causing violent, delusional insanity—the product of a scientist working at a secret military research establishment:

He took an organism known as mycoplasma and mutated it ... so that if it entered the bloodstream it would attack the healthy existing cells and travel as a parasite to the brain. [22]

As dangerous as the mutated organism is, it accidentally gets loose from the laboratory—instantly transforming anyone who encounters it into a crazed maniac. That provides Herbert with all the excuse he needs to churn out an outrageous horror story—as the book's back cover blurb summarizes:

In an exclusive private school, students sexually assaulted and mutilated their teachers, then savagely turned on one another. . . . At a seaside resort, thousands of people joined in a monstrous act of self-destruction. . . . In the city, mass copulation and insane slaying spread. [23]

The Fog represents the dystopian extreme of CBW. But, all things being possible in fiction, what about the other end of the spectrum? What if a chemical agent could make people better human beings, rather than worse ones?

The idea of a global utopia, in which everyone lives in peace and harmony, has been around for a long time. The reason it's never materialized is that it requires everyone in the world, down to the last individual, to put the welfare of the community above their own self-interest. That's fine as far as the proponents of such utopias are concerned—they tend to be instinctive altruists—but what about everyone else?

What's needed is a magic chemical that turns everyone into the same kind of altruist as the utopian dreamers. Just such a chemical crops up in one of Agatha Christie's last novels, *Passenger to Frankfurt* (1970), which she wrote at the age of 80. As an elderly female character, possibly modelled on Christie herself, says to a research scientist:

> Instead of inventing all these germ warfares and these nasty gases, and all the rest of it, why don't you just invent something that makes people feel happy? [24]

The scientist takes the suggestion seriously, and invents a substance he names Benvo, for benevolence. Its effects are described by his research assistant in the following terms:

> "There is something that can change a man in his outlook to life—his reaction to people and to life generally. He may be in a state of homicidal fury, he may be pathologically violent, and yet, by the influence of Project Benvo, he turns into something, or rather someone, quite different. He becomes—there is only one word for it, I believe, which is embodied in its name—he becomes benevolent. He wishes to benefit others. He exudes kindness. He has a horror of causing pain or inflicting violence. Benvo can be released over a big area, it can affect hundreds, thousands of people if manufactured in big enough quantities, and if distributed successfully."
>   "How long does it last?" said Colonel Munro. "Twenty-four hours? Longer?"
>   "You don't understand," said Miss Neumann. "It is permanent." [25]

In Christie's novel, Project Benvo is only fully revealed in the final chapter—a kind of *deus ex machina* to rescue a world rapidly dissolving into anarchy. To see how it might actually work, it's necessary to turn to another novel of the same period: John Brunner's *The Stone that Never Came Down* (1973).

Portraying a similar vision of a world descending into chaos, Brunner introduces his miracle cure—a self-replicating, empathy-enhancing drug called VC, or "Viral Coefficient" [26]—near the start of the narrative, not the end. That leaves him with the rest of the book to show how, in effect, giving the world's population an injection of "niceness" would solve all its problems. Whether that's a credible proposition, or simply wishful thinking, is left as an exercise for the reader.

Hearts and Minds

By far the oldest form of "government mind control" is propaganda, whether aimed at the enemy—by leaflet dropping, for example—or at one's own population. The latter type is the most insidious, and George Orwell lampooned it in his novel *Nineteen Eighty-Four* (1949). It's the job of the protagonist, Winston Smith, to generate such propaganda on behalf of the inaptly named Ministry of Truth.

Since the present book is supposed to be about science, and not politics or sociology, it would be straying off-topic to go into any detail on the subject of Cold War propaganda. It is, however, worth making a brief detour to look at the role played by science fiction in such propaganda.

It has already been mentioned—in the first chapter, "The Super-Bomb"—that Hollywood often used sci-fi films to make political points, from the overt anti-nuclear message of *The Day the Earth Stood Still* (1951) to the more subtly implied anti-communism of *The Invasion of the Body Snatchers* (1955).

On the other side of the Iron Curtain, Soviet space propaganda often drew on sci-fi imagery, illustrating real-world space missions with artwork that looked more like something from a 1930s pulp magazine (see Fig. 3).

In the West, propagandizing SF stories were fairly evenly divided between the political left and right. This was particularly true after the United States became involved in the highly divisive Vietnam War. But did that have anything to do with the Cold War?

To some, the answer was obviously "yes". North Vietnam was a communist state, fighting with weapons and equipment that had been supplied to them by the Soviet Union. To others, however, the connection to the Cold War was a tenuous one at best. Viewed in a wider context, the war was part of Vietnam's long struggle for independence against European colonialism, and the involvement of Russia was an incidental issue. According to this latter view, the United States—whose own origins lay in just such a struggle—was fighting on the wrong side.

The Vietnam War divided America down the middle—and it divided the science fiction community, too. In 1967, a number of SF magazines printed a paid advertisement displaying the names of 72 writers who "believe the United States must remain in Vietnam to fulfill its responsibilities to the people of that country", alongside those of another 82 who "oppose the participation of the United States in the war in Vietnam" [27].

The "pro-war" side included five of the best-known names in SF at that time: Poul Anderson, John W. Campbell, Robert A. Heinlein, Larry Niven

Fig. 3 A Soviet postage stamp commemorating Gherman Titov's spaceflight in August 1961—with his actual Vostok spacecraft replaced by an archetypal sci-fi rocketship (public domain image)

and Jerry Pournelle. Of the remaining names on that side, six are mentioned elsewhere in this book: David A. Kyle, Alva Rogers, Fred Saberhagen, George O. Smith, G. Harry Stine and Jack Williamson.

On the other hand, the "anti-war" side boasted five big-name authors of its own: Isaac Asimov, Ray Bradbury, Philip K. Dick, Harlan Ellison and Ursula K. Le Guin. And once again, there are another six names that feature elsewhere in this book: Forrest J. Ackerman, Anthony Boucher, Lester del Rey, Mack Reynolds, Robert Silverberg and Donald A. Wollheim.

In other words, the two lists were pretty evenly balanced.

One of the less well-known authors on the anti-war list, James Sallis, went on to edit an anthology of Vietnam-inspired science fiction called *The War Book* (1969). As he says in the introduction:

> In projecting our own dilemmas onto a much larger, more abstract background, the technique of genre SF leads almost inevitably to a preoccupation with war. . ..
> It occurred to me that much of the best current science fiction was dealing in specific terms with this theme: was examining the moral problems brought home to us all now in Vietnam. [28]

As true as this was, stories of this type can't really be classed as propaganda. The audience for written science fiction was too small—and arguably too discerning and perceptive—for any amount of pro- or anti-Vietnam proselytizing to have an impact on the way people thought. It was a different matter, though, when it came to television.

This brings us to another famous name on that "anti-war" list—one that wasn't mentioned earlier because it doesn't really qualify, belonging as it does not to an author of printed SF but a TV producer. This, of course, was Gene Roddenberry—the creator of *Star Trek*. Although the series was set in the 23rd century, many of the plotlines from Roddenberry's original series—which aired from 1966 to 1969—can be seen as overt commentaries on the then-ongoing Cold War.

Consider, for example, the 23rd episode of the second season, broadcast in March 1968. In the words of genre historian James van Hise:

"Omega Glory" features yet another parallel history: the warring Kohms and Yangs parallel the communists and Yankees of the Vietnam War era. A starship captain has set himself up as a warlord with the Kohms; Kirk and Spock finally rally the Yangs when Kirk realizes that their sacred words are actually a distortion of the preamble to the United States Constitution! This powerful episode was written entirely by Roddenberry himself, an unusual occurrence at the time. [29]

One of the most popular characters in that original run of *Star Trek* was the half-human, half-Vulcan Mr Spock. Roddenberry often used Spock as a mouthpiece for a worldview—which he referred to as "the Vulcan philosophy of universal brotherhood" [30]—that closely mirrored his own.

Another *Star Trek* character that Roddenberry used to didactic effect was Ensign Chekov, who joined the cast at the start of the second season in 1967. The significant thing about Chekov is that he was Russian; Roddenberry wanted to emphasize the transient nature of the US-Soviet antagonism dominating world affairs at the time. Amusingly, however, a rumour grew up that the Russians themselves had been behind the introduction of Chekov, as James van Hise recounts:

A press release (later revealed to have exaggerated the truth by falsifying the incident) claimed that the show was criticized by the Russian communist newspaper Pravda for, among other things, its lack of a Russian character in the Enterprise's otherwise multi-national crew. And so . . . Roddenberry reportedly created the character of Ensign Pavel Chekov, a young officer with a heavy accent, to satisfy Soviet angst. [31]

War Is Peace

"War is Peace" was one of the slogans of the Ministry of Truth in *Nineteen Eighty-Four*. In that novel, it was an example of "doublethink . . . the power of holding two contradictory beliefs in one's mind simultaneously, and accepting both of them" [32].

During the Cold War, on the other hand, the war/peace ambiguity was much more literal. It was a war in name, and each side put a large part of its economy into the production of weapons to defeat the other side. On the other hand, there wasn't even a brief moment of actual fighting between Russia and America. Political leaders in both countries used the word "peace" far more than "war", both in their dealings with each other and when talking to their own people.

In the United States, this ambivalence reached its peak in Project Ploughshare (or "Plowshare", as it is spelled in American English). The name comes from the Bible, where Chapter 2, verse 4 of the book of Isaiah says:

> They will hammer their swords into ploughshares and their spears into sickles. Nation will not lift sword against nation, no longer will they learn how to make war. [33]

A ploughshare is a bladed agricultural implement used for tilling soil, so the image of fabricating one from a no-longer-needed sword makes sense. But could the same thing be done with more advanced weapons?

Philip K. Dick satirized this idea in his 1967 novel *The Zap Gun*, which features the "ploughsharing" of a whole variety of high-tech weapons into useful domestic appliances. Before it appeared in book form, the novel was serialized in magazine form under the title "Project Plowshare" (see Fig. 4).

The ludicrousness of adapting a specialized modern weapon for peaceful use is highlighted in a quotation at the very start of *The Zap Gun*:

> The guidance system of weapons-item 207, which consists of 600 miniaturized electronic components, can best be ploughshared as a lacquered ceramic owl which appears to the unenlightened only as an ornament; the informed knowing, however, that the owl's head, when removed, reveals a hollow body in which cigars or pencils can be stored. [34]

If anything, the real-world Project Ploughshare was even crazier than Dick's novel, concentrating as it did on the seemingly impossible task of finding

Fig. 4 An illustration from the January 1966 issue of *Worlds of Tomorrow*, depicting Philip K. Dick's story "Project Plowshare"—later published in book form as *The Zap Gun* (public domain image)

peaceful uses for thermonuclear weapons. At first sight, that might be taken to mean more carefully controlled nuclear technology, such as the use of nuclear reactors for power generation. That's far too sensible, however, and Project Ploughshare was actually thinking in terms of all-out kiloton- or megaton-scale nuclear explosions.

One of its key proponents was none other than Edward Teller. He's cropped up time and again in this book—first as the father of the H-bomb, then as a model for Stanley Kubrick's Dr Strangelove, then as the intellectual power behind Ronald Reagan's Strategic Defence Initiative. Project Ploughshare, however, cast him in what may be his most way-out role of all. As historian Audra J. Wolfe explains:

> Between 1957 and 1975, Teller and his colleagues spent hundreds of millions of dollars devising plans to use nuclear devices as convenient tools for mining operations, oil and gas exploration, and most famously, earth-moving projects. Bombs might be used to create a new Alaska harbour or, perhaps, a new Panama Canal. [35]

Fortunately for posterity—think of all that radiation!—very few of Project Ploughshare's ideas were tested in the real world. One that was is described here by military historian David Baker:

> On 10 December 1967 an underground nuclear detonation took place at a site in New Mexico . . . as an exercise in "fracking", the extraction of natural gas from

rock formations. Adjacent to existing gas wells, the team drilled to a depth of 1.3 km and lowered a 4 m by 46 cm diameter nuclear device that was detonated with a yield of 29 kT, almost twice that of the bomb dropped on Hiroshima. There was great enthusiasm for this method of fracking and engineers calculated that the explosion would provide a cavernous void within the gas-bearing sandstone into which the gas would escape, from where it could be piped to the surface … In 1969 some eight million cubic metres of gas was extracted, which was found to be contaminated with tritium. Two further nuclear fracking tests were carried out in Colorado. …. While considerable quantities of natural gas were liberated, none of it was useable due to contamination. [36]

From a science fiction perspective, fracking may seem an unimaginatively mundane application of an atom bomb. That can't be said of Project Orion—which, if it had been built, would have given the world a bomb-powered spaceship.

The nuclear-powered rocket was quite a common SF trope in the 1940s and 1950s—but apart from the name, it had little in common with a nuclear bomb. Instead, the idea—a perfectly valid one—was to use a controlled nuclear reactor, of the kind that powers a nuclear submarine, to heat gases which would then be ejected through an exhaust nozzle like a conventional rocket. Project Orion was nothing like that.

Euphemistically referred to as "nuclear pulse propulsion", it was quite literally powered by atom bombs—hundreds or even thousands of them. These would have been exploded one after another against a pusher plate to provide the necessary forward momentum (see Fig. 5).

Set up in 1958, Project Orion was ambitious, to say the least. To quote science writer Brian Clegg:

> Orion would have weighed in at about 10,000 tons. It was designed to carry up to 150 people with a payload capacity of thousands of tons. …. Unthinkably from a modern viewpoint, despite the requirement to fling out a string of nuclear bombs behind its 40-metre pusher plate, the original plan was that Orion would take off from the Earth's surface, making use of a nuclear test site to lift off through the atmosphere on its stream of explosions. The fission devices used to propel it would start with 0.1 kiloton devices every second at take-off, building up to 20 kilotons every 10 seconds. Six times a minute it would set off devices more powerful than the Hiroshima bomb. [37]

With its incessant, kiloton-scale explosions—and the associated hazards of deadly radiation and transistor-destroying electromagnetic pulse—Project Orion posed engineering challenges on an unprecedented scale. We'll never

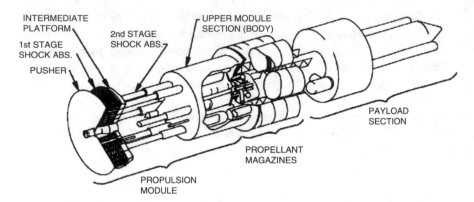

Fig. 5 Hypothetical configuration of the Project Orion spacecraft, powered by exploding nuclear bombs against a "pusher plate" (NASA image)

know if they could have been overcome, because the project was cancelled in 1963 in the wake of the Partial Nuclear Test Ban Treaty—which restricted peacetime nuclear detonations to those carried out underground (such as the fracking tests mentioned earlier).

When the previously secret Project Orion became public knowledge, it was briefly considered by Arthur C. Clarke and Stanley Kubrick for inclusion in *2001: A Space Odyssey* (1968), which they were developing in parallel as a novel and a film. The latter would have been a follow-up to Kubrick's *Dr Strangelove*, which had carried the subtitle "How I Learned to Stop Worrying and Love the Bomb". As Clarke recalled later:

> When we started work on *2001*, some of the Orion documents had just been declassified, and were passed on to us . . . but after a week or so Stanley decided that putt-putting away from Earth at the rate of 20 atom bombs per minute was just a little too comic. Moreover—recalling the finale of *Dr Strangelove*—it might seem to a good many people that he had started to live up to his own title and had really learned to Love the Bomb. [38]

Encouraging people to "love the bomb"—or at any rate to be less terrified by it—was, of course, the real motive behind the superficially crazy ideas of Project Ploughshare and Project Orion. In other words, they were a form of propaganda in themselves—an attempt to counter the vocal "ban the bomb" movement that was thriving at exactly the same time.

Information and Disinformation

While hypnosis, drugs and propaganda may be the most obvious ways of playing with an enemy's mind, there's another much more subtle tactic. You can confuse the enemy so much he doesn't know which way is up. The undisputed master of this technique—in a science-fictional context, at least—was Eric Frank Russell. At the peak of his popularity in the mid-1950s, he produced several variations on the theme for *Astounding* magazine—including "Diabologic" (March 1955), "Plus X" (June 1956) and "Nuisance Value" (January 1957).

Russell later expanded "Plus X" into a novel called *Next of Kin* (1959). The protagonist is a resourceful human who is taken prisoner by aliens—and then proceeds to run circles around them. In the episode taken from the "Plus X" short story, he seeks to confuse his captors by fabricating a piece of—totally spurious—technology called a bopamagilvie:

> He took his half-length of wire, broke it into two unequal parts, shaped the shorter piece to form a neat loop with two legs each three or four inches long. He tried to make the loop as near to a perfect circle as possible. The longer piece he wound tightly around the loop so that it formed a close-fitting coil with legs matching the others. . ... Lastly he counted the number of turns to the coil. There were 27. It was important to remember these details because in all likelihood he would have to make a second gadget as nearly identical as possible. That very similarity would help to bother the enemy. When a plotter makes two mysterious objects to all intents and purposes the same, it is hard to resist the notion that he knows what he is doing and has a sinister purpose. [39]

When questioned about the bopamagilvie, he claims it's a device for contacting an invisible being called a Eustace. There's no such thing, of course, and the real purpose of the device is to tie up the enemy's attention to the exclusion of all else. In this, it's completely successful (see Fig. 6):

> They were interested in one subject and one only, namely, bopamagilvies. It seemed that they'd been playing for hours with his samples, had achieved nothing except some practice in acting daft, and were not happy about it. On what principle did a bopamagilvie work? Did it focus telepathic output into a narrow, long-range beam? At what distance did his Eustace get beyond range of straight conversation and have to be summoned with the aid of a gadget? Why was it necessary to make directional search before obtaining a reply? How did he know how to make a coiled-loop in the first place? [40]

Fig. 6 In Eric Frank Russell's novel *Next of Kin*, the human protagonist confuses the enemy with a bogus piece of technology called a bopamagilvie. This illustration comes from a shorter version, "Plus X" in the June 1956 issue of *Astounding Science Fiction* (public domain image)

Russell's bopamagilvie is a perfect example of disinformation. This has always been a factor in war, traditionally in the context of strategy—witness, for example, the huge efforts made during World War Two to confuse the Nazis as to the location of the D-Day landings. In the Cold War, however—as in Russell's story—the art of disinformation was applied to the field of science and technology.

The whole Cold War, in fact, can be considered a kind of "virtual conflict", with scientific and technological one-upmanship serving as a proxy for actual fighting. Against this background, it was imperative to control what the other side knew or believed about one's own capabilities.

The most obvious consequence of this was the paranoid secrecy of the Cold War, aimed at preventing the enemy from learning about the various technologies in your possession. Equally important, however, was persuading them that you also possessed technologies which you didn't. That's where disinformation comes in.

It's all about sowing confusion—which means anything goes, as long as the other side doesn't find out where you really stand. As Mark Pilkington wrote in his book *Mirage Men* (2010):

When it came to secret warfare—espionage and counter-espionage, intelligence and counter-intelligence, psychological operations, disinformation and covert action—nothing was true and everything was permitted, just as long as nobody found out. [41]

The most valuable commodity of the Cold War was certainty, and the most damaging thing you could inflict on the enemy was uncertainty. One way to do that was through bogus science. One possible example of this was described in the previous chapter, "Star Wars". To repeat a quote from scientist Gerold Yonas, who worked on Ronald Reagan's Strategic Defence Initiative: "I don't know any technical person who believed all of the stuff we said" [42].

That's not to say that SDI—with its talk of space-based X-ray laser weapons—was out-and-out nonsense, but there's no doubt it exaggerated the potential benefits and underplayed the problems and risks. If America's budget-planners had looked at the proposals objectively, they would never have given them the green light. They were simply too expensive, too poorly thought-out and too impractical.

There was another consideration, however. By allowing the research to go ahead, its mere existence served a militarily useful purpose—by confusing the Soviets into thinking the Americans knew something they didn't. That forced them to try to catch up, wasting valuable time and resources in the process. In that sense, SDI was the real-world equivalent of Eric Frank Russell's bopamagilvie.

Similar thinking may have underpinned other bizarre areas of Cold War research, such as anti-gravity. That's a subject with a venerable SF pedigree, as the *Encyclopedia of Science Fiction* explains:

The idea of somehow counteracting gravity is one of the great SF dreams: it is gravity that kept us Earthbound for so long, and even now the energy expenditure required to escape the gravity well of Earth or any other massive celestial body is the main factor that makes spaceflight so difficult and expensive. [43]

As the entry goes on to say, anti-gravity was already a well-established SF trope in 1901, when H. G. Wells produced *The First Men in the Moon*—with its miraculous, and completely fictitious, gravity-shielding material called Cavorite.

Regardless of its science-fictional heritage, anti-gravity research was something the US government was happy to pump money into during Cold War—albeit most often under the more respectable-sounding designation of "gravity control". In 1956, for example, a body calling itself the Gravity

Research Group produced a report called *Electrogravitics Systems* for the US Air Force. Reviewing the status of anti-gravity work by various commercial companies, the report stated that:

> Glenn Martin say gravity control could be achieved in six years, but they add it would entail a Manhattan Project type of effort to bring it about. .... General Electric is working on electronic rigs designed to make adjustments to gravity . . . Bell also has an experimental rig intended, as the company puts it, to cancel out gravity. . ... Lear Inc, autopilot and electronic engineers, have a division of the company working on gravity research and so also has the Sperry division of Sperry-Rand. This list embraces most of the US aircraft industry. The remainder, Curtis-Wright, Lockheed, Boeing and North American have not yet declared themselves, but all these four are known to be in various stages of study. [44]

That report dates from the early years of the Cold War—but interest in the subject continued to the very end. As aerospace journalist Nick Cook relates:

> The US Air Force ... proclaimed its interest in the field with a document, published in August 1990, called the "Electric Propulsion Study". Its objective was to "outline physical methods to test theories of inductive coupling between electromagnetic and gravitational forces to determine the feasibility of such methods as they apply to space propulsion." Stripped of the gobbledegook, it was really asking whether there was any theory out there that might permit the engineering of an anti-gravity device. [45]

Despite numerous documents of this type finding their way into the public domain, there isn't a shred of evidence that any practical application ever came of them. That suggests that, even if these studies weren't out-and-out disinformation, the US government was happy to spend money on exotic-sounding research that had very little chance of success. No doubt the aim, just as with SDI, was to keep the Soviets guessing.

The same might be said of several other areas of "weird science", including UFOs, extra-sensory perception and super-powered soldiers. There's enough material there to fill a whole chapter—and that's what's coming up next.

References

1. I. Kuhn, E.A. Woolf, J. Willard (screenplay), *The Hand of Fu Manchu* (Metro-Goldwyn-Mayer, 1932)
2. TV Tropes, Hypno Ray, http://tvtropes.org/pmwiki/pmwiki.php/Main/HypnoRay
3. F. Nadis, *The Man from Mars* (Penguin, New York, 2014), p. 66
4. D.H. Childress, *Lost Continents & the Hollow Earth* (Adventures Unlimited, Illinois, 1999), pp. 83, 84
5. F. Nadis, *The Man from Mars* (Penguin, New York, 2014), p. 101
6. R. Palmer, in *Amazing Stories* (August 1946), p. 6
7. F. Nadis, *The Man from Mars* (Penguin, New York, 2014), p. 110
8. A. Balthazar, *Project MK-Ultra and Mind Control Technology* (Adventures Unlimited, Illinois, 2017), p. 76
9. S. Rohmer, *The Si-Fan Mysteries* (Corgi, London, 1967), p. 136
10. D. Guyatt, Police State of Mind, in *Fortean Times* (February 1997), pp. 32–39
11. T. O'Toole, Moscow Microwaves: No Harm Seen, in *Washington Post* (21 November 1978), https://www.washingtonpost.com/archive/politics/1978/11/21/moscow-microwaves-no-harm-seen/7a4b045f-e2ff-401e-a1f5-34e748d4cf13/
12. F. Kaplan, Something in the Air, in *Slate* (September 2017), http://www.slate.com/articles/news_and_politics/war_stories/2017/09/the_cold_war_incident_that_could_hold_clues_about_what_s_happening_to_u.html
13. Geneva Protocol to Hague Convention (1925), https://en.wikisource.org/wiki/Geneva_Protocol_to_Hague_Convention
14. A. MacLean, *The Satan Bug* (Fontana, Glasgow, 1964), pp. 39, 40
15. A. Balthazar, *Project MK-Ultra and Mind Control Technology* (Adventures Unlimited, Illinois, 2017), p. 24
16. A. Balthazar, *Project MK-Ultra and Mind Control Technology* (Adventures Unlimited, Illinois, 2017), pp. 21–31
17. A. Balthazar, *Project MK-Ultra and Mind Control Technology* (Adventures Unlimited, Illinois, 2017), pp. 55, 56
18. A. Balthazar, *Project MK-Ultra and Mind Control Technology* (Adventures Unlimited, Illinois, 2017), p. 67
19. A. Roberts, Reservoir Drugs, in *Fortean Times* (May 2010), pp. 38–42
20. B. Aldiss, *Barefoot in the Head* (Corgi, London, 1971), p. 17
21. B. Aldiss, *Barefoot in the Head* (Corgi, London, 1971), pp. 19, 20
22. J. Herbert, *The Fog* (New English Library, London, 1975), pp. 148, 149
23. J. Herbert, *The Fog* (New English Library, London, 1975), back cover copy
24. A. Christie, *Passenger to Frankfurt* (Fontana, London, 1972), p. 168
25. A. Christie, *Passenger to Frankfurt* (Fontana, London, 1972), pp. 182, 183

26. J. Brunner, *The Stone that Never Came Down* (New English Library, London, 1978), p. 56
27. M. Ashley, *The Illustrated Book of Science Fiction Lists* (Virgin Books, London, 1982), pp. 162, 163
28. J. Sallis, *The War Book* (Panther, London, 1971), p. 12
29. J. van Hise, *The Unauthorized History of Trek* (Harper Collins, London, 1997), p. 46
30. S.E. Whitfield, G. Roddenberry, *The Making of Star Trek* (Ballantine, New York, 1968), p. 226
31. J. van Hise, *The Unauthorized History of Trek* (Harper Collins, London, 1997), pp. 38, 39
32. G. Orwell, *Nineteen Eighty-Four* (online text), http://orwell.ru/library/novels/1984/english/
33. *The New Jerusalem Bible* (Darton, Longman & Todd, London, 1990), p. 882
34. P.K. Dick, *The Zap Gun* (Panther, London, 1975), p. 5
35. A.J. Wolfe, *Competing with the Soviets* (Johns Hopkins University Press, Baltimore, 2013), p. 1
36. D. Baker, *Nuclear Weapons* (Haynes, Yeovil, 2017), p. 65
37. B. Clegg, *Final Frontier* (St Martin's Press, New York, 2014), p. 209
38. A.C. Clarke, *The Lost Worlds of 2001* (Sidgwick & Jackson, London, 1972), p. 125
39. E.F. Russell, *Next of Kin* (Mandarin, London, 1989), pp. 116, 117
40. E.F. Russell, *Next of Kin* (Mandarin, London, 1989), pp. 174, 175
41. M. Pilkington, *Mirage Men* (Constable, London, 2010), p. 72
42. N. Hey, *The Star Wars Enigma* (Potomac Books, Washington DC, 2007), p. 210
43. Encyclopedia of Science Fiction, Antigravity, http://www.sf-encyclopedia.com/entry/antigravity
44. N. Cook, *The Hunt for Zero Point* (Arrow, London, 2002), pp. 27, 28
45. N. Cook, *The Hunt for Zero Point* (Arrow, London, 2002), p. 120

Weird Science

In which that most archetypal of all sci-fi themes, the arrival of visitors from another planet, spills over into the real world—as the start of the Cold War coincides with a sudden upsurge in sightings of unidentified flying objects. Were they extraterrestrial spacecraft, top-secret spyplanes, an exercise in disinformation or just another symptom of Cold War paranoia? Whatever the case, UFOs weren't the only "weird science" development that looked like it had been plucked straight from the pages of SF. The Cold War also saw serious attempts being made to exploit extra-sensory perception and other superhuman powers to military ends.

Flying Saucers

Earlier chapters have shown how the Cold War appropriated numerous subjects that had previously been confined to science fiction, from robotics and space travel to beam weapons and mind control rays. For SF, however, such topics were often peripheral to its main preoccupation—the notion of alien beings from other worlds beyond Earth. Even this, however, became a topic of serious speculation during the Cold War.

People have been seeing mysterious objects in the sky for centuries, and even speculating that they might be intelligent visitors from outer space. Despite that, it was only at the outset of the Cold War that humanity really began its all-out obsession with unidentified flying objects, or UFOs. The

© Springer International Publishing AG, part of Springer Nature 2018
A. May, *Rockets and Ray Guns: The Sci-Fi Science of the Cold War*, Science and Fiction,
https://doi.org/10.1007/978-3-319-89830-8_6

media coined the term "flying saucer" in 1947—and almost immediately it was a fixture in headlines around the world.

The timing wasn't just a coincidence. Almost everyone agrees that the upsurge in UFO sightings was tied to the wider events of the Cold War—although there are different theories as to why that was. Maybe they really were alien spacecraft, suddenly attracted to Earth by our atomic tests and space rockets. Maybe they were Soviet secret weapons, or American ones. Or maybe the UFO phenomenon was just another example of the kind of technological disinformation described in the previous chapter, "Mind Games".

Probably the best-known of all UFO events occurred in June 1947, when a supposed flying saucer crashed in Roswell, New Mexico. Something crashed, anyway. Although an early press release did indeed describe it as a "flying disc", it was soon dismissed by the authorities as a weather balloon. Since then, armchair ufologists have come up with their own pet theories. "The fact is that there are more than a dozen explanations for what happened", as conspiracy researcher Nick Redfern puts it [1]. In the present context, however, the various theories are of less interest than the location—Roswell itself.

As described in the first chapter, "The Super-Bomb", the atom bombs dropped on Hiroshima and Nagasaki were delivered by aircraft of the US Army Air Force's 509th Composite Group. At that time—August 1945—they were based on Tinian Island in the Pacific. Just three months later, however, the 509th moved to a new base in the mainland United States: the Roswell Army Air Field. That meant that, when a mysterious something crashed in June 1947, it did so less than a hundred kilometres from the only military site on the planet capable of delivering nuclear weapons.

To ufologists, the inference was obvious: the crashed object must have been an alien spacecraft sent to observe the latest developments on Earth. To sceptics, it was equally obvious that it had been a secret military project—or perhaps an advanced Soviet spy-plane.

It may come as a surprise to modern readers, but in the very earliest days of the "flying saucer" phenomenon it was this last view that caused the most consternation. As author and social scientist David Clarke explains:

> During the Cold War the military intelligence agencies in the US and UK were naturally interested in any "unidentified" objects—such as aircraft, missiles and satellites—that might have a hammer and sickle painted on their fuselage. [2]

And to quote another social historian, S. D. Tucker:

In 1948, even US Naval Intelligence preferred to speculate that the saucers were some strange form of Soviet psychological warfare, aimed at fooling America into thinking that their precious atom bombs were not the ultimate weapons they had presumed them to be. [3]

Even among the general public, there was little interest in the "extraterrestrial hypothesis" at the start. Here is David Clarke on the results of a Gallup poll conducted in 1947, shortly after the first appearance of flying saucers:

When asked "what do you think these saucers are?" a third said they had no idea. Another third believed they were products of "imagination, optical illusion, mirage etc". Ten per cent believed the whole story was a hoax, while others mentioned American secret weapons or "weather forecasting devices". Conspicuous by its absence from the poll was any hint of belief in the saucers as craft from other worlds. [4]

One of the first people to popularize an extraterrestrial interpretation of UFOs was the author Donald Keyhoe. Most ufological sources emphasize his early career as a pilot with the US Marine Corps—but arguably more relevant is the fact that he also spent several years as a prolific writer of pulp fiction. Some of this even fell into the category of science fiction, such as the story "The Master of Doom" in the May 1927 issue of *Weird Tales*. Its "mad scientist" villain is planning to destroy the world with a magnetic super-weapon, as he explains to the story's hero:

Have you ever thought of what would happen if all the iron in the world suddenly had its polarity completely reversed? ... You would not, of course. You are like all the thick-headed scientists who laughed at my early discoveries. In 24 hours they will have laughed for the last time. [5]

With the arrival of the UFO phenomenon, Keyhoe turned his writing talents to non-fiction—and his first book on the subject, *The Flying Saucers Are Real* (1950), became an instant best-seller (see Fig. 1).

Keyhoe's books promoted two attention-grabbing ideas: that UFOs are extraterrestrial spacecraft, and that the US Air Force[1] knows a lot more about them than it lets on. Quickly becoming articles of faith among UFO believers, this changed the whole complexion of the subject. To quote David Clarke and Andy Roberts: "Keyhoe's writings had a massive impact on public

---

[1] Keyhoe had no great affection for the Air Force, having been a pilot in a rival service—the Marine Corps.

Fig. 1 The cover of Donald Keyhoe's book *The Flying Saucers Are Real* (1950)—one of the first non-fiction works to portray UFOs as alien spacecraft (public domain image)

opinion and as a result he became a major headache for the USAF for years to come" [6].

Like many other UFO writers, Keyhoe scoured historical records for evidence that the phenomenon had been going on for centuries. He contended, however, that it was no coincidence that the current wave of sightings correlated so closely with the onset of the Cold War:

> According to the reports, Europe, the most populated area, had been more closely observed than the rest of the globe until about 1870. By this time, the United States, beginning to rival Europe in industrial progress, had evidently become of interest to the spaceship crews. From then on, Europe and the western hemisphere, chiefly North America, shared the observers' attention. . .. World War Two had drawn more attention, and this had obviously increased from 1947 up to the present time. Our atomic bomb explosions and the V-2 high-altitude experiments might be only coincidence, but I could think of no other development that might seriously concern dwellers on other planets. [7]

Ironically, this last possibility was also given brief consideration by Keyhoe's arch-nemesis, the US Air Force. In a report dated April 1949, they speculated on the possibility that UFOs came from the planet Mars:

Martians have kept a long-term routine watch on Earth, and have been alarmed by the sight of our A-bomb shots as evidence that we are warlike and on the threshold of space travel. The first flying objects were spotted in the spring of 1947, after a total of five atomic explosions. Of these, the first two were in positions to be seen from Mars. It is likely that Martian astronomers, with their thin atmosphere, could build telescopes big enough to see A-bomb explosions on Earth. [8]

In the realm of science fiction, the idea of worried aliens being drawn to Earth by our atomic tests provides the plot of *The Day the Earth Stood Still* (1951). As the movie's benevolent alien Klaatu explains:

For our policemen, we created a race of robots. Their function is to patrol the planets—in spaceships like this one—and preserve the peace. In matters of aggression, we have given them absolute power over us; this power cannot be revoked. At the first sign of violence, they act automatically against the aggressor. The penalty for provoking their action is too terrible to risk. The result is that we live in peace, without arms or armies, secure in the knowledge that we are free from aggression and war. [9]

In the early 1950s, a number of individuals—collectively known as "contactees"—claimed to have received direct communications from alien visitors, usually in the form of anti-nuclear warnings. There was an obvious similarity here to *The Day the Earth Stood Still*—"a storyline many contactees of the day ripped off shamelessly" as S. D. Tucker puts it [10]. He goes on to point out that, in those days, there was a fine line between articulating anti-nuclear sentiments and being seen as pro-communist:

Many contactees were mocked, ridiculed and even feared by mainstream 1950s society not only as kooks and loons, but as possible communists and dangers to society, too. Such fears were not wholly unfounded. [10]

The most famous of the contactees was a middle-aged resident of California named George Adamski. Previously a guru-style teacher of eastern mysticism and philosophy, he ascribed similarly progressive views to the extraterrestrials. After attracting the attention of the FBI, he explained the situation to one of their agents in the following way:

If you ask me, they probably have a communist form of government. That is a thing of the future—more advanced. . ... Russia will dominate the world and we will then have an era of peace for a thousand years. . ... The United States today is in the same state of deterioration as was the Roman Empire prior to its collapse and it will fall just as the Roman Empire did. The government in this country is a corrupt form of government and capitalists are enslaving the poor. [11]

In retrospect, it seems pretty obvious that Adamski and his fellow contactees simply wanted to promote their own political views—and believed they could reach out to a larger audience by attributing those views to visitors from outer space. For most people, however, the effect was the opposite of the intended one. The contactees simply succeeded in destroying their own credibility—and with it, that of the whole subject of UFOs.

From a modern reader's perspective, some of the claims made by contactees in the 1950s were so ludicrous it raises the suspicion that they were government plants, deliberately discrediting the UFO subject in a carefully planned disinformation campaign. As Mark Pilkington says in his book *Mirage Men* (2010):

Whether they were CIA puppets or not, Adamski and the other contactees . . . ensured that it would be some time before the UFO subject would again be taken seriously. [11]

That said, it wasn't in the CIA's interest to debunk UFO sightings completely—not when they might be exploited in the Cold War battle of nerves against the Soviets. Pilkington's book includes the following quote from a report addressed to the director of the CIA in September 1952:

The question, therefore, arises as to whether or not these sightings . . . could be used from a psychological warfare point of view, either offensively or defensively. [12]

The same report goes on to conclude that:

A study should be instituted to determine what, if any, utilization could be made of these phenomena by United States psychological warfare planners and what, if any, defences should be planned in anticipation of Soviet attempts to utilize them. [12]

How far did the CIA go with such studies? Perhaps not as far as some people believe, says Pilkington—his own conclusion being that the CIA exploited the UFO subject in an occasional, opportunistic way rather than a systematic one:

I have suggested that popular ideas about UFOs have been shaped and manipulated by disinformation specialists within America's intelligence apparatus: the Mirage Men. However ... I don't think that there is any grand plan in the perpetuation of the UFO myth. The lore is perfectly capable of maintaining itself; it has a vibrant life of its own, supported by a complex patchwork of believers, promoters, seekers and charlatans, and nourished by the sightings of thousands of new witnesses every year. The lore does, however, provide a useful cover for certain clandestine operations, and is thus employed by the Mirage Men if and when it is expedient to do so. [13]

Whether or not Pilkington's "mirage men" existed in reality, there's no doubt amateur ufologists generated some pretty useful disinformation themselves—albeit unwittingly. A case in point is Area 51—a high-security research establishment hidden deep in the Nevada desert. Perhaps more than any other Cold War military site, the Americans wanted to keep the goings-on there secret from the Soviets. But what was the nature of those goings-on?

One person who knew the truth was engineer Edward Lovick, who visited the site on many occasions when he was working on the design of a stealthy, super-fast successor to the U-2 spyplane called the A-12. His involvement began in late 1959, when, in his own words, "a new, and much better, facility was established at Groom Lake in Area 51 of the Atomic Energy Commission reservation north of Las Vegas, Nevada" [14].

The site was initially used to measure the aircraft's radar-reflecting characteristics while it was on the ground. As Lovick explains, this was later expanded to include airborne testing:

In April of 1960, a flight test facility for measuring radar echoes dynamically was started at Groom Lake in Area 51 in Nevada. .... During the latter part of 1963 and January 1964, several flights of an A-12 were performed over the dynamic flight test range using the ground radars at Groom Lake. The radar echoing characteristics were measured while the aircraft flew under cruising conditions. [15]

There was clearly a lot going on that had to be concealed from the prying eyes of the Soviets: the existence of the new spyplane, its physical appearance, its flying capabilities—and most importantly of all, the specific radar frequencies the Americans were testing. When ufologists started telling the world, over and over at the top of their voices, that Area 51 was where the Americans kept their captured alien spacecraft—well, that was a perfect cover story as far as the US Air Force was concerned.

Towards the end of the Cold War, in the 1980s, a new stealth aircraft was developed and test-flown at Area 51: the bizarre-looking Lockheed F-117A.

Fig. 2 Seen nose-on like this, the Lockheed F-117A stealth fighter bears a close resemblance to the classic image of a flying saucer (public domain image)

What would an outside observer—particularly one obsessed with UFOs—have made of it? From certain angles it looks a lot like a classic flying saucer (see Fig. 2).

There was another—and more sinister—way in which Cold War governments could exploit the popular belief in UFOs. That was to cover up incidents that might otherwise have caused serious political embarrassment. One such situation is recounted by Jenny Randles in a book called *The UFOs that Never Were* (2000).

The incident in question occurred on 11 January 1973, in the skies over the village of Chilton in the United Kingdom—or more specifically, over the village primary school. Around 9 o'clock that morning, dozens of children saw a glowing orange ball hovering over the school. By a lucky chance, the same scene was witnessed by an ex-serviceman, named Peter Day, who happened to have a movie camera to hand—so he was able to snatch a short film of the object.

Like most of the schoolchildren, Day thought at first that he'd seen a UFO. He soon learned, however, that an F-111 fighter plane of the US Air Force had crashed that morning about 50 km to the north-east. Assuming the fireball he had seen must have been connected with the crash, Day asked the USAF if they wanted to see the film he'd made.

Surprisingly, they turned the offer down—and, in the words of Jenny Randles, "gave the witness every indication that what he had filmed was not connected with their aerial mishap" [16]. Day went on to make the same offer to the British Ministry of Defence—with the same result. They didn't seem to think the glowing ball he'd witnessed had any bearing on the F-111 crash.

As far as Day was concerned, this was tantamount to official confirmation that what had filmed was a genuine UFO. His short video clip became established in the UFO community as one of the best pieces of evidence for the reality of the phenomenon. In his 1991 *UFO Encyclopedia*, for example, John Spencer says it "remains today one of the most important pieces of film available to UFO researchers" [17].

Yet it was that very year, 1991, that the shocking truth emerged—as a result of a query under the US Freedom of Information Act. Despite all official denials, it turned out there really was a connection with the F-111 incident. When the crew first got into trouble, at 9 am over the village of Chilton, they decided they needed to vent excess fuel before attempting an emergency landing. Following standard procedure, they used the aircraft's afterburner to burn the fuel in the air after it had been jettisoned, to prevent it contaminating the ground.

Unfortunately, they chose to do this directly over a primary school. As Jenny Randles wrote in 2000:

It is unlikely that either the MoD or USAF would have been very keen to trigger a public relations disaster by having to admit this fact. . .. Even to this day, 27 years later, there have been no public revelations about just how close Chilton School may have come to disaster on that January morning. . . .. Perhaps it was felt better that this story remain a UFO legend than the quite terrifying reality be made clear. [18]

It's important to note there was no actual "disinformation" on the part of the authorities; they just sat back and let the UFO community speculate. To quote Jenny Randles again:

In seeking to prove an alien reality or to defend the unexplained nature of cases such as this, ufologists may well serve as unwitting puppets to the powers-that-be. [18]

Psychic Spies

In a scene towards the end of Robert A. Heinlein's novel *Starship Troopers* (1959), a specially talented individual called a "spatial senser" walks around a landscape creating a map of the enemy tunnels beneath [19]. The image is reminiscent of the ancient art of "dowsing" for underground water or mineral deposits. There's no scientific justification for this practice, and it's usually lumped with other psychic abilities such as telepathy. In spite of this, efforts really have been made to exploit such talents in a military context—hunting for buried landmines, for example. Here is David Clarke on the subject, drawing on documents stored in Britain's National Archives:

> During the Cold War, the threat to personnel grew worse as new enemies developed deadly non-metallic mines in theatres of conflict such as the Middle East. . ... The problem had become so serious by 1968 that the army's Chief Scientist asked the Military Engineering Experimental Establishment (MEXE) to investigate unconventional methods of mine detection. Scientists from MEXE devised a series of experiments to test dowsers' powers in the field. [20]

The results of these controlled tests were "no more reliable than a series of guesses". Nevertheless, there is anecdotal evidence that some dowsers are consistently more successful than this. It's possible, of course, that they're simply using subliminal cues from their normal senses. Even Heinlein's protagonist "wondered if that 'special talent' was simply a man with incredibly acute hearing" [21].

Other supposed psychic powers are harder to explain away. Heinlein explored some of them—and their possible utility in a military context—in an earlier story, "Project Nightmare" (1953). As one of the characters explains:

> Extrasensory perception, or ESP, is a tag for little-known phenomena—telepathy, clairvoyance, clairaudience, precognition, telekinesis. They exist; we can measure them; we know that some people are thus gifted. But we don't know how they work. The British, in India during World War One, found that secrets were being stolen by telepathy. . ... It is possible that a spy 500 miles away is now "listening in"—and picking your brains of top secret data. [22]

Despite the reference to WW1, the story is set in its own Cold War present. An experiment is set up to find out if a group of psychically talented individuals can make a subcritical mass of plutonium go critical simply by using ESP. As the organizer puts it:

> Today we will find out if the mind can change the rate of neutron emission in plutonium. . ... By standard theory, theory that works, that subcritical mass out there is no more likely to explode than a pumpkin. Our test group will try to change that. They will concentrate, try to increase the probability of neutrons escaping, and thus set off that sphere as an atom bomb. [22]

Fig. 3 Illustration for Robert A. Heinlein's story "Project Nightmare", from *Amazing Stories*, April 1953 (public domain image)

The experiment is successful—but the group's power is really put to the test when they're called on to deal with a genuine crisis. In a mirror image of the test scenario, they have locate and neutralize a series of atom bombs that Soviet agents have planted in American cities (see Fig. 3).

During the 1950s, one person who became increasingly enthusiastic about the idea of psychic powers was John W. Campbell, the editor of *Astounding Science Fiction*. Preferring the more scientific sounding name of "psionics", he printed as many stories on the subject as he could get hold of. The culmination of these—in a military context, at least—was Murray Leinster's "Short History of World War Three", from the January 1958 issue of *Astounding*.

After recounting a series of apparently inexplicable defeats suffered by the Soviet Union, the story winds up as follows:

This was the conclusion of World War Three. It wasn't at all dramatic. The Russians simply gave up. The rest of the world could bomb them, and they couldn't bomb back. The rest of the world could spot their subs, and they couldn't spot back. ... So ultimately the Russians had to decide to coexist or be smashed—if they'd be allowed to coexist.

They were. They got quite peaceful, after a few dozen years. They even quit being communists. ... So World War Three passed into oblivion instead of history, and the Russians didn't even know that it had been fought, and they'd lost it, until a long, long time later, when the Historical Section of the Navy put out a volume entitled "History of the Psionic Devices Division of the United States Naval Research Laboratory." It didn't make much of a stir, at that. Most people take psi for granted, like electricity, and don't try to go too deeply into it. It's too bad. It's had its effect on history. World War Three was a psionic war, instead of an atomic one. [23]

As we've seen throughout this book, no matter how weird science fiction got, Cold War reality was prepared to try and emulate it. ESP was yet another sci-fi favourite that really did attract government attention. As with so much of the "advanced research" of the time, it's difficult to know whether this was driven by genuine scientific curiosity, a cynical desire to confuse the opposition—or some combination of the two.

Even the origins of Cold War ESP research are shrouded in confusion. As Jim Marrs explains in his book *Psi Spies* (2007):

The entire affair began as a response to a perceived "psychic gap" between the United States and the Soviet Union—and this gap was generated by an apparent hoax. The saga began in February 1960, when the French magazine *Science et Vie* published an article reporting that an American nuclear submarine, the USS Nautilus, had conducted successful telepathy experiments. ... Years later the story, vehemently denied by the US Navy at the time, was alleged to be a hoax perpetrated by a French writer who later sold a book on the subject. ... Hoax or not, parapsychologists in the Soviet Union used the story to good advantage. [24]

The Soviet response, the year after the original French article appeared, was to set up a special parapsychology laboratory at the University of Leningrad. Why would they do that, if they weren't on to something? Even if the Americans hadn't been interested in the subject up to this point, they were now. The East-West "psychic arms race" was well and truly on.

In hindsight, the situation was farcical. That comes across to good effect in one of the flashback scenes in the 2009 comedy film *The Men Who Stare at Goats*. A military intelligence specialist, Brigadier General Hopgood, is briefing two senior Pentagon officials. While the Hopgood character is fictitious, the salient details remain true-to-life:

Official: But why did the Soviets begin this type of research?

Hopgood: Well sir, it looks like they heard about our attempt to telepathically communicate with one of our nuclear subs, the Nautilus, while it was under the polar cap.

Official: What attempt?

Hopgood: There was no attempt, sir. It seems the story was a French hoax. But the Russians think the story about the story being a French hoax is just a story sir.

Second official: So, they've started psi research because they thought we were doing psi research, when in fact we weren't doing psi research?

Hopgood: Yes sir. But now that they are doing psi research, we're going to have to do psi research, sir. We can't afford to have the Russians leading the field in the paranormal. [25]

In another scene from the movie, General Hopgood is seen reading a paperback book called *Psychic Discoveries Behind the Iron Curtain*. That too is historically accurate: the book was written by Sheila Ostrander and Lynn Schroeder and published in 1970. Among other things, it states that:

All research on ESP in the USSR is, of course, ultimately financed by the government. There is every indication from multiple sources that psi research with military potential is well financed by the Soviet Army, Secret Police and other paramilitary agencies. [26]

As mentioned in the previous chapter, "Mind Games", the main purpose of disinformation is to keep the other side guessing. Whether or not the Soviet ESP research was deliberate disinformation, it certainly did that. Quoting from a 1972 CIA report:

Soviet efforts in the field of psi research, sooner or later, might enable them to do some of the following: (a) Know the contents of top secret US documents, the movements of our troops and ships and the location and nature of our military installations. (b) Mould the thoughts of key US military and civilian leaders at a distance. (c) Cause the instant death of any US official at a distance. (d) Disable, at a distance, US military equipment of all types, including spacecraft. [27]

In the light of such worries, it's not surprising the US put efforts of its own into developing and exploiting psychic abilities. Around the time that CIA report was written, two American scientists—Harold Puthoff and Russell Targ at the Stanford Research Institute in California—were experimenting with a new form of ESP they called "remote viewing", or RV. This started out as purely civilian research, and in 1974 Puthoff and Targ had a paper published in the prestigious scientific journal *Nature* claiming success for the technique:

> A channel exists whereby information about a remote location can be obtained by means of an as yet unidentified perceptual modality. [28]

In a typical RV session, the viewer was given a target location—for example a set of geographic coordinates—and asked to close their eyes and visualize it. They then produced a pencil sketch of what they saw. The results were often impressive—or at least, the results that Puthoff and Targ chose to publish were (see Fig. 4).

As far as the CIA was concerned, RV looked too good to ignore. They promptly snapped up Puthoff, Targ and their whole team of remote viewers for covert espionage work. The journalist Nick Cook, who interviewed Puthoff after the work had been declassified, summarized the situation as follows:

> For the next decade and a half, US "psychic spies" roamed the Soviet Union, using nothing more than the power of their minds to reconnoitre some of the Russians' most secret research and development establishments. . . . Puthoff showed me a picture that one of his team of remote viewers had drawn of an "unidentified research centre at Semipalatinsk", which had been deeply involved in Soviet nuclear weapons work. He then showed me some declassified artwork of the same installation, presumably drawn from a US spy satellite photograph. The two were damn near identical. [29]

Although the reality or otherwise of psychic phenomena has been debated for centuries, almost everyone agrees that—at best—they are capricious. It was no different with RV, which was never as consistent as its government sponsors would have liked. In the words of one of the most successful remote viewers, Ingo Swann:

> The average accuracy of a spontaneous remote viewing is about 20 per cent at most . . . and 20 percent accuracy is not competitive with other intelligence-gathering methods. [30]

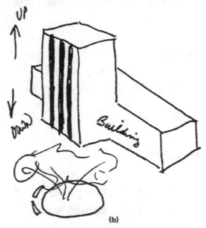

Fig. 4 Example of (a) a remote viewing target location and (b) the sketch of it produced by a remote viewer, from a Stanford Research Institute report on "Special Orientation Techniques" published in March 1980 (public domain image)

That's not to say RV was never employed during the Cold War, but its results were always treated with a pinch of salt. As Jim Marrs recounts:

> Unit members stressed that information produced by the psi spies ... was never used by itself, but only to double-check other intelligence sources or to point to an avenue of investigation by other sources. [31]

In spite of that, the Soviets took the threat from American remote viewers seriously, making efforts of their own to duplicate and counteract that threat. Quoting Marrs again:

> The Soviet KGB laboriously screened more than a million people in an effort to locate "super naturals", persons with the greatest amount of psychic power. These super psychics became the Soviet Union's psi spies, sometimes assigned to seek out their Western counterparts. [32]

These real-world events of the 1970s and 1980s are starting to sound like all those "psionics" stories John W. Campbell published in *Astounding Science Fiction* in the 1950s. To take just one example, consider the novel *Brain Twister* by Randall Garrett and Laurence M. Janifer, which Campbell serialized in 1959. It deals with a "psychic counterspy" scenario of just the kind described by Jim Marrs. Here's the back-cover blurb from the 1992 paperback edition:

> FBI agent Malone has a problem. Someone is reading the minds of the nation's top scientists, stealing the most vital secrets from the closely guarded new space-drive project. Now agent Malone must find another telepath who can catch the spy. [33]

Mutants and Monsters

In science fiction, the existence of psychic powers such as telepathy is often associated with genetic mutation of the human species. During the Cold War, this was almost always portrayed as a consequence of atomic radiation.

One of the first stories to deal with radiation-spawned mutants actually appeared several months before the Hiroshima bomb. This was "The Piper's Son" by Henry Kuttner, from the February 1945 issue of *Astounding*. It's set after a large-scale nuclear war, in which "the hard radiations brought about one true mutation: hairless telepaths" [34]. These telepathic mutants are referred to as "Baldies", and Kuttner went on to write a whole series of stories about them.

Although the Baldies are the only mutation that proved viable, they weren't the only one to appear—as one of them explains:

> Some mighty queer specimens came out of the radioactive-affected areas around the bomb targets. Funny things happened to the germ plasm. Most of 'em died out; they couldn't reproduce; but you'll still find a few creatures in sanatoriums—two heads, you know. And so on. [34]

Two-headed mutants became just as much a sci-fi cliché as telepathic ones. Both types—and many others—began to appear in large numbers in the late 1940s and early 1950s. As Robert Silverberg says, in the introduction to his 1974 anthology *Mutants*:

The atomic bombs that brought World War Two to a sudden and frightful close had, as might be expected, a powerful impact on the imaginations of science fiction writers. For years after Hiroshima and Nagasaki, the pages of science fiction magazines were filled with stories describing in grim detail the horrifying human mutations that would come into the world after the atomic devastation of World War Three. There were so many of these stories that readers wearied of them and editors stopped buying them. [35]

One author who returned time and again to the mutant theme—in the 1950s and later—was Philip K. Dick. Not all his thoughts on the subject were as depressingly downbeat as Silberberg's remarks suggest. Take 1953s "Planet for Transients", for example. In this story, Dick portrays mutations—not just human ones, but at all levels in the animal and plant kingdoms—as a natural evolutionary adaptation to the now-radioactive environment of Earth's surface. While ordinary humans are forced to wear lead-lined suits whenever they go outside, the mutants thrive:

These altered progeny littered the Earth. A crawling, teeming, glowing horde of radiation-saturated beings. In this world, only those forms which could use hot soil and breathe particle-laden air survived. Insects and animals and men who could live in a world with a surface so alive that it glowed at night. [36]

Another of Dick's "mutant" stories—and one that's particularly noteworthy in the Cold War context—is "The Hood Maker", from 1955. Like the Baldies in Kuttner's stories, the "teeps" in this story are mind-reading mutants. The twist, however, is that they are also the ultimate McCarthyites—using their powers to root out communist sympathizers and other disloyal citizens:

Before the teeps, loyalty probes had been haphazard. Oaths, examinations, wire-tappings, were not enough. . .. The problem, apparently insoluble, had found its answer in the Madagascar Blast of 2004. Waves of hard radiation had lapped over several thousand troops stationed in the area. Of those who lived, few produced subsequent progeny. But of the several hundred children born to the survivors of the blast, many showed neural characteristics of a radically new kind. A human mutant had come into being—for the first time in thousands of years. The teeps appeared by accident. But they solved the most pressing problem the Free Union faced: the detection and punishment of disloyalty. [37]

Sadly, the real-world effects of radiation tend to be fatal long before they can lead to mutations, beneficial or otherwise. In 1961, for example, an accident on board the Soviet nuclear submarine K-19 killed most of its crew. When the reactor began to overheat, the captain, Yuri Posetiev, was forced to send a team of eight men into the reactor room to deal with it. To quote Sherry Sontag and Christopher Drew, from their account of the incident in *Blind Man's Bluff* (2000):

They remained in the compartment for two hours, braving the heat and the invisible particles that shot through their bodies. Each received one hundred times the lethal dose of radiation. [38]

The reactor was eventually repaired, but at a terrible cost. As Sontag and Drew go on to say:

The team of eight, those first men into the compartment during the crisis, died before the week was out. They were buried in lead coffins. Posetiev lingered longer: three weeks. Other crewmen who had come too close to the outer door of the reactor compartment lasted a month, some a little longer, before they too succumbed. [38]

To most people at the time, the fate of the K-19 crew would have come as no surprise at all. The fear of radiation, and the creeping but inevitable death it brought, was a major source of paranoia throughout the Cold War. Much of this angst focused on the radioactive "fallout" that followed a nuclear explosion. From the early 1950s, public fallout shelters began to appear across America, while smaller versions were advertised that families could install in their own basements (see Fig. 5).

The fallout shelter craze was satirized by Philip K. Dick in his short story "Foster, You're Dead" (1955), in which these shelters are portrayed as desirable consumer goods. Here is a scene showing the young protagonist mesmerized by one in a showroom:

He gazed at the shelter for a long time. It was mostly a big tank, with a neck at one end that was the descent tube, and an emergency escape-hatch at the other. It was completely self-contained; a miniature world that supplied its own light, heat, air, water, medicines and almost inexhaustible food. When fully stocked there were visual and audio tapes, entertainment, beds, chairs, vidscreen,

TEMPORARY BASEMENT FALLOUT SHELTER

Fig. 5 Depiction of a domestic fallout shelter, from a US government publication circa 1957 (public domain image)

everything that made up the above-surface home. It was, actually, a home below the ground. Nothing was missing that might be needed or enjoyed. A family would be safe, even comfortable, during the most severe H-bomb and bacterial-spray attack. [39]

Philip K. Dick, of course, represents the highbrow end of 1950s science fiction. At the other extreme, the entertainment industry cashed in on the public's fear of radiation by portraying its effects in exaggerated form. This was the era of "creature features"—often involving monsters created by one type of nuclear mishap or another. One of Hollywood's first forays into this genre, *Them!* (1954), was described by *Time Out* several decades later in the following terms:

By far the best of the '50s cycle of creature features, *Them!* and its story of a nest of giant radioactive ants (the result of an atomic test in the New Mexico desert) retains a good part of its power today. All the prime ingredients of the total mobilization movie are here ... intermixed with gloomy speculation about the effect of radioactivity. [40]

Also dating from 1954 is the Japanese movie *Godzilla*—the titular monster of which is an ancient dinosaur-like creature brought back to life by H-bomb testing. Godzilla, of course, went on to become one of the world's best-known fictional monsters, up there with Frankenstein's famous creation. When it first appeared, *Godzilla* was just as symbolic of the hopes and fears of its time as Mary Shelley's *Frankenstein* had been in the previous century. As Maria J. Pérez Cuervo wrote in *Fortean Times*:

> The Cold War and the nuclear menace called for a different kind of escapism. The fictional creatures born in this context had a mythical quality, but with a contemporary twist. This was provided by radiation, which could awaken ancient beasts, as in *Godzilla*. ... At the end of most of these films, different nations unite to restore the lost peace; wishful thinking in uneasy times. [41]

Super-Soldiers

Just as space travel is the archetypal sci-fi subject, so superheroes are the archetypal comic-book subject. The military subgenre of SF may be focused on space warfare, but its comic-book counterpart centres on the super-soldier.

The first of the breed was Captain America, who made his debut in March 1941, just a few months before the United States entered World War Two. Originally a weedy young man named Steve Rogers, he gained his powers from a serum developed as part of a secret government project. With a ready-made enemy to fight—in the form of the Nazis—Captain America became one of the most popular superheroes of the WW2 years.

After the end of war, however, comic-book tastes changed and his adventures were discontinued. A brief attempt to revive the character was made in 1953—reinvented for the Cold War as "Captain America, the Commie Smasher". Would readers agree that, as Steve Rogers was made to say, "communists are the Nazis of the 1950s" [42]? Unfortunately they didn't—and the revived Captain America was dropped after nine months.

A second revival in 1964 was much more successful. Under the guiding inspiration of editor Stan Lee, the company—now called Marvel Comics—was busy creating a whole new generation of superheroes. The wartime Captain America (no mention was made of his commie-smashing interlude) was slotted into the new continuity by claiming he'd been frozen in a block of ice since 1945. Found in this state by Marvel's best-known super-team, the Avengers, he was duly thawed out and invited to join. Both Captain America and the Avengers are, of course, still going strong in the 21st century—courtesy of the Marvel Cinematic Universe (MCU).

It may come as a surprise to the MCU's newer fans, but some of its best-known characters can trace their roots back to the darkest days of the Cold War. Take that other famous Avenger, Iron Man, for example—alias millionaire businessman Tony Stark. Many of his earliest adventures, in the pages of *Tales of Suspense*, saw him pitted against communist adversaries like the Red Barbarian, the Crimson Dynamo and the Black Widow (the last-named—an early incarnation of the character played by Scarlett Johansson in the MCU films—originally being a Soviet spy).

Even Iron Man's own origin, as related in the March 1963 issue of *Tales of Suspense*, was inextricably tied to the Cold War. To quote comic-book historian Sean Howe:

Wounded and kidnapped by Wong-Chu, the "Red Guerrilla Tyrant", Stark is ordered to develop a weapon for the communist enemy. Instead he constructs a metal suit that will keep his failing heart in operation, and also serve as armour in which he can escape. [43]

Iron Man's metal suit—which makes him just as much a "super-soldier" as Captain America—is functionally reminiscent of the powered armour that featured a few years earlier in Heinlein's *Starship Troopers*. In the latter case, the armoured exoskeleton uses servo motors and feedback circuits to amplify the wearer's natural movements—allowing them, for example, to jump much longer distances than they normally could.

Iron Man, on the other hand, can fly. There's an echo here of another long-established sci-fi cliché: the jet-pack (see Fig. 6).

Here is what the *Encyclopedia of Science Fiction* has to say on the subject:

The use of jet-packs for individual atmospheric flight is a venerable SF tradition. An early appearance is in the first tale of Buck Rogers in the 25th century, "Armageddon 2419 AD" (August 1928 *Amazing*) by Philip Francis Nowlan, where US soldiers in a future war fly by means of rockets strapped to their backs. [44]

It's an appealing idea—and not just to sci-fi fans. In the late 1950s, the US Army provided the Bell Aircraft Corporation with funding to develop a "rocket belt". As the name suggests, this was literally a small rocket motor strapped to the user's back. It had a serious shortcoming, though. Like any rocket, it burnt fuel at a tremendous rate—a problem made worse by the fact that the fuel tank had to be small and light enough for a person to carry.

Bell managed to make a working rocket belt, but it was little practical use because the maximum flight time was only a minute or so. The army, as a

Fig. 6 The cover of the May 1954 issue of *Science Fiction Quarterly*, depicting an archetypal sci-fi jet-pack (public domain image)

result, was singularly unimpressed. It was a different matter when it came to the general public, as Brian Clegg explains:

> After a successful demonstration of the belt at the 1964 New York World's Fair, the makers of the James Bond movies got in touch. They wanted something new and dramatic for the iconic pre-title chase sequence for their latest film. In *Thunderball*, Bond was going to escape his attackers using a rocket belt. . . .. The appearance in the movie was a great success, even though many assumed that the rocket belt was a special effect. Despite many test flights at Bell, the military once more lost interest because little progress was being made toward practical battlefield technology. Rocket belts would continue to be used for publicity stunts and as entertainment—notably in the opening ceremony of

the 1984 Los Angeles Olympic Games … but they weren't practical transport devices. [45]

Returning to the subject of Marvel superheroes: many of Stan Lee's creations of the early 1960s had their origins in that favourite Cold War bugbear, radiation. The first of the new comics to appear was *The Fantastic Four*, in November 1961. To put that date in context: the Soviets had launched Yuri Gagarin into orbit in April that year, followed by Gherman Titov in August, while the United States could only match that with two suborbital flights, by Alan Shepard in May and Gus Grissom in July. The first orbital flight by an American—John Glenn in February 1962—still lay in the future.

Against that background, the first issue of *The Fantastic Four* sees scientist Reed Richards planning a private-venture spaceflight of his own, together with test pilot Ben Grimm and friends Sue and Johnny Storm. The critical role played by the Cold War comes across in the dialogue:

Ben Grimm: If you want to fly to the stars, then you pilot the ship. Count me out! You know we haven't done enough research into the effect of cosmic rays. They might kill us all out in space!

Sue Storm: Ben, we've got to take that chance—unless we want the commies to beat us to it! [46]

The flight goes ahead, but the spacecraft's poor shielding means the crew experiences the full force of the cosmic rays Ben Grimm was worried about. Improbably from a real-world perspective, but perfectly logically in comic-book terms, all four of them gain super-powers as a result.

The radiation that turned the Fantastic Four into superheroes was natural in origin. Applying comic-book logic, however, a trick that works with one type of radiation will work just as well with another. So for his next creation, Stan Lee turned to the Cold War's number one obsession: the atom bomb. The result, in May 1962, was the first issue of *The Incredible Hulk*—described by Sean Howe as "a nuclear-age updating of the Dr Jekyll/Mr Hyde story":

Again the scientific frontiers of the Cold War were vital to the story: Dr Bruce Banner was preparing to test a Gamma Bomb for the US military when a reckless teenager named Rick Jones drove his convertible onto the desert testing site on a dare. Banner called for a delay on the test while he got Jones to safety, but a communist spy on the lab team proceeded anyway, bombarding Banner with radiation. [47]

Exposure to the bomb's radiation transforms Banner into a super-strong, green-skinned giant. As with all Lee's radiation-spawned superheroes (and super-villains for that matter), the physiological effects seem to be fairly random—but ultimately always beneficial to the recipient. Next in line for the treatment, in the August 1962 issue of *Amazing Fantasy*, was a bullied young high-school student named Peter Parker. Here is Stan Lee's textual commentary as Parker watches a scientific demonstration:

> A few minutes later Peter Parker forgets the taunts of his classmates as he is transported to another world—the fascinating world of atomic science. But as the experiment begins, no one notices a tiny spider, descending from the ceiling on an almost invisible strand of web. .... Accidentally absorbing a fantastic amount of radioactivity, the dying insect [sic], in sudden shock, bites the nearest living thing at the split second before life ebbs from its radioactive body. [48]

The person bitten is, of course, Peter Parker himself—who is thus transformed into Marvel's latest superhero, Spider-Man. His earliest adventures were illustrated by Steve Ditko, who also collaborated with Stan Lee on another, completely different character. This was Doctor Strange, "the master of the mystic arts", who debuted in the July 1963 issue of *Strange Tales*.

Capable of feats like levitation, astral projection and psychokinesis, Doctor Strange was just as powerful, in his own way, as any other Marvel superhero. The difference was that these powers came not from the world of science—via a chemical serum or radiation accident—but that of mysticism. He learned them from "the Ancient One"—a wise old teacher he encountered in the Himalayas.

Belief in mystical abilities of this type really does feature in certain Eastern traditions. To quote from a Buddhist manual from the 12th century, for example:

> Supernormal powers include the ability to display multiple forms of one's body, to appear and vanish at will, to pass through walls unhindered, to dive in and out of the Earth, to walk on water, to travel through the air. ... The knowledge of others' minds is the ability to read the thoughts of others and to know directly their states of mind. ... The divine eye is the capacity for clairvoyance, which enables one to see heavenly and or earthly events, both far or near. [49]

Buddhism is a famously pacifist religion, so it seems somewhat sacrilegious to point out that powers of this type might be very useful from a military point of view. Nevertheless, such powers are discussed in exactly that context in the film *The Men Who Stare at Goats* (2009), mentioned earlier in chapter (the title, for example, comes from the idea that an adept could kill a goat simply by staring at it).

In a flashback sequence to the time of the Cold War, one of the movie's characters—an Army officer named Bill Django—suggests that these and other mystical talents should be cultivated by the US military:

> We must become the first superpower to develop super-powers. We must create warrior monks—men and women who can fall in love with everyone, sense plant auras, pass through walls ... and see into the future. [25]

Although the movie is a work of fiction, *The Men Who Stare at Goats* was loosely based on a non-fiction book of the same title, written by Jon Ronson in 2004. The real-world counterpart of Bill Django was Lieutenant Colonel Jim Channon—who, in 1979, produced a serious proposal for the US Army called *The First Earth Battalion*. While it does use terms like "warrior monks", the proposal is nothing like as kooky as the film suggests. It's not about developing literal super-powers, so much as drawing on non-western traditions—such as martial arts and meditation—which are already employed by fighting forces in other parts of the world. Here is Channon himself, writing in *The Guardian* in 2009:

> The Battalion mythology I developed was a creative thinking tool designed to encourage the young leaders in the army to think of new ways, with the aim of changing the nature of war and improving the chances of survival for all involved. It was intended to stretch the imagination. [50]

Nevertheless, *The First Earth Battalion* manual does contain its fair share of mystical mumbo-jumbo, almost worthy of Doctor Strange himself. For example, Fig. 7 depicts the "guerrilla guru" of the future, who employs "eastern philosophy", the new age" and "the occult" alongside "space-age technology" and "the traditional" [51].

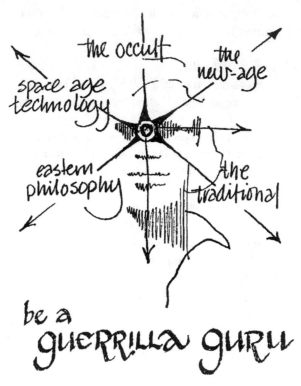

Fig. 7 A page from Jim Channon's *First Earth Battalion* manual from 1979, in which he proposed that soldiers of the future should adopt a more mystical philosophy (public domain image)

References

1. N. Redfern, *The Roswell UFO Conspiracy* (Lisa Hagan Books, Bracey, VA, 2017), p. 23
2. D. Clarke, *UFO Drawings from the National Archives* (Four Corners, London, 2017), p. 8
3. S.D. Tucker, *Space Oddities* (Amberley, Stroud, 2017), p. 153
4. D. Clarke, *How UFOs Conquered the World* (Aurum Press, London, 2015), p. 38
5. D. Keyhoe, The Master of Doom, in *Weird Tales* (May 1927), pp. 581–600
6. D. Clarke, A. Roberts, *Flying Saucerers* (Alternative Albion, Loughborough, 2007), p. 19
7. D. Keyhoe, *The Flying Saucers are Real* (Project Gutenberg, 2004), http://www.gutenberg.org/files/5883/5883-h/5883-h.htm
8. S.D. Tucker, *Space Oddities* (Amberley, Stroud, 2017), p. 152
9. E.H. North (screenplay), *The Day the Earth Stood Still* (Twentieth Century Fox, 1951)

10. S.D. Tucker, *Space Oddities* (Amberley, Stroud, 2017), pp. 146, 147
11. M. Pilkington, *Mirage Men* (Constable, London, 2010), p. 105
12. M. Pilkington, *Mirage Men* (Constable, London, 2010), p. 84
13. M. Pilkington, *Mirage Men* (Constable, London, 2010), pp. 296, 297
14. E. Lovick, *Radar Man* (iUniverse, New York, 2010), p. 128
15. E. Lovick, *Radar Man* (iUniverse, New York, 2010), p. 157
16. J. Randles et al., *The UFOs that Never Were* (London House, London, 2000), p. 60
17. J. Spencer, *The UFO Encyclopaedia* (Headline, London, 1991), p. 92
18. J. Randles et al., *The UFOs that Never Were* (London House, London, 2000), pp. 74, 75
19. R.A. Heinlein, *Starship Troopers* (Hodder, London, 2015), pp. 245–247
20. D. Clarke, *Britain's X-traordinary Files* (Bloomsbury, London, 2014), p. 95
21. R.A. Heinlein, *Starship Troopers* (Hodder, London, 2015), p. 252
22. R.A. Heinlein, Project Nightmare, in *14 Great Tales of ESP*, ed. by I.P. Stone (Coronet, London, 1970), pp. 111–136
23. M. Leinster, Short History of World War Three, in *Astounding Science Fiction* (October 1958, UK edition), pp. 107–118
24. J. Marrs, *Psi Spies* (New Page Books, Franklin Lakes NJ, 2007), pp. 95–96
25. P. Straughan (screenplay), *The Men Who Stare at Goats* (Momentum Pictures, 2009)
26. J. Marrs, *Psi Spies* (New Page Books, Franklin Lakes, NJ, 2007), pp. 97, 98
27. J. Marrs, *Psi Spies* (New Page Books, Franklin Lakes NJ, 2007), p. 101
28. D. Clarke, *Britain's X-traordinary Files* (Bloomsbury, London, 2014), p. 111
29. N. Cook, *The Hunt for Zero Point* (Arrow, London, 2002), p. 162
30. J. Marrs, *Psi Spies* (New Page Books, Franklin Lakes NJ, 2007), p. 153
31. J. Marrs, *Psi Spies* (New Page Books, Franklin Lakes NJ, 2007), p. 208
32. J. Marrs, *Psi Spies* (New Page Books, Franklin Lakes NJ, 2007), p. 166
33. R. Garrett, L.M. Janifer, *Brain Twister* (Carroll & Graf, New York, 1992), back cover copy
34. H. Kuttner, *Mutant* (Mayflower, London, 1962), pp. 7–32
35. R. Silverberg, *Mutants* (Corgi, London, 1977), p. 9
36. P.K. Dick, Planet for Transients, in *A Handful of Darkness* (Panther, London, 1980), pp. 46–60
37. P.K. Dick, The Hood Maker, in *Philip K. Dick's Electric Dreams* (Gollancz, London, 2017), pp. 117–134
38. S. Sontag, C. Drew, *Blind Man's Bluff* (Arrow, London, 2000), pp. 286, 287
39. P.K. Dick, Foster, You're Dead, in *Philip K. Dick's Electric Dreams* (Gollancz, London, 2017), pp. 138–162
40. T. Milne (ed.), *The Time Out Film Guide* (Penguin, London, 1989), p. 595
41. M.J. Pérez Cuervo, The Politics of Monsters, in *Fortean Times* (Christmas 2017), pp. 30–37

42. C. Moss, Captain America, McCarthyite, *The Atlantic* (April 2014), https://www.theatlantic.com/entertainment/archive/2014/04/captain-america-mccarthyite/360183/

43. S. Howe, *Marvel Comics: The Untold Story* (Harper Perennial, New York, 2012), p. 43

44. Encyclopedia of Science Fiction, Flying, http://www.sf-encyclopedia.com/entry/flying

45. B. Clegg, *Ten Billion Tomorrows* (St Martin's Press, New York, 2015), pp. 92, 93

46. S. Lee, *Origins of Marvel Comics* (Simon & Schuster, New York, 1974), p. 28

47. S. Howe, *Marvel Comics: The Untold Story* (Harper Perennial, New York, 2012), p. 39

48. S. Lee, *Origins of Marvel Comics* (Simon & Schuster, New York, 1974), p. 142

49. B. Bodhi (ed.), *A Comprehensive Manual of Abhidhamma* (Buddhist Publication Society, Sri Lanka, 1993), p. 344

50. J. Channon, My First Earth Battalion Comes to Life in The Men Who Stare at Goats, *The Guardian* (November 2009), https://www.theguardian.com/film/2009/nov/02/men-who-stare-at-goats1

51. J. Channon, The First Earth Battalion (1979), https://archive.org/stream/FirstEarthBattalionManual/First_Earth_Battalion_Manual

Future Shock

In which the Cold War comes to an end, somewhat sooner and less dramatically than most science fiction writers had expected. Nevertheless, SF got many things right, and by the end of the Cold War the world took the genre—with its vision of a future defined by scientific and technological progress—far more seriously than it had at the start.

Welcome to the Future

People have always speculated about the future. Before the Cold War, however, such speculations were more likely to focus on politics and social change than science and technology. In the 1920s and 1930s, following the communist revolution in Russia, most people could foresee a future clash between the communist East and the capitalist West—but few of them imagined it would revolve around atomic super-weapons and space travel. Science fiction, on the other hand, dealt with such topics on a routine basis (see Fig. 1).

When people talk about "the future" today, it's usually taken for granted that said future will be defined by its technology. Yet that's a view that only gradually emerged over the course of the Cold War. One of its pioneers—ahead of the trend as ever—was Arthur C. Clarke. One of his first non-fiction books, *Profiles of the Future*, was published in 1962. Subtitled "enquiry into the limits of the possible", it focused squarely on a future that would be driven by technological advances. To quote the cover blurb from the first edition:

© Springer International Publishing AG, part of Springer Nature 2018
A. May, *Rockets and Ray Guns: The Sci-Fi Science of the Cold War*, Science and Fiction,
https://doi.org/10.1007/978-3-319-89830-8_7

Fig. 1 The cover of *Science Wonder Quarterly*, Spring 1930. Prior to the Cold War, the idea that the future would primarily be defined by technological advancements was rarely discussed outside SF magazines of this type (public domain image)

Round the world in 80 minutes—Fuel-less flight—Travel by conveyor belt and hovercraft—Overcoming gravity . . . Communication across space—Our problems on Jupiter, Mercury, Venus—Electronic brains . . . etc. [1]

Just over a decade later, in 1973, Clarke produced a second edition of *Profiles of the Future*. By this time, technological marvels that would have seemed like science fiction at the time of the first edition were making everyday headlines. A dozen Apollo astronauts had walked on the Moon, the Vietnam War had seen the use of autonomous guided missiles, a pre-production version of the Concorde airliner had flown at twice the speed of sound, and geosynchronous satellites made worldwide TV broadcasting possible. But there was an even bigger change, as Clarke pointed out in his preface to the new edition:

What has changed … is our entire attitude towards the future, and especially towards technology as a whole. *Profiles of the Future* was one of the first samples of a deluge of books about the future; today, there are societies, foundations, journals devoted to the study of "futuristics". [2]

Not only did the Cold War change the world's attitude to the future, it changed its attitude to science as well. It was no longer an amusing sideline pursued by eccentric inventors and absent-minded professors. In the West, the turning point was the Sputnik crisis of October 1957, which persuaded the American public—and its politicians—that science was a subject that had to be taken seriously.

The Soviets took science seriously—and it seemed they were better at it than the Americans. As science fiction writer Lester del Rey put it in April 1958: "Russia is working harder at the job of producing future scientists and engineers than we are" [3].

Another SF writer, Mack Reynolds, dramatized the situation in a 1960 short story called "Combat". As the protagonist, an American secret agent, says early on in the story:

The best men our universities could turn out went into advertising, show business and sales—while the best men the Russians and Chinese could turn out were going into science and industry. … The height of achievement over there is to be elected to the Academy of Sciences. Our young people call scientists eggheads, and their height of achievement is to become a TV singer or a movie star. [4]

This being science fiction, an alien spacecraft then arrives on Earth. To be specific, it lands in Moscow—which comes as a surprise to the Americans, who always assumed alien visitors would choose to land in the world's most advanced nation. That, of course, is exactly what they did do—as one of them later explains to the protagonist:

What you should have done was try to excel Russian science, technology and industry. Had you done that you might have continued to be the world's leading nation. [4]

In the real world, the American government took the Sputnik crisis seriously. As early as September 1958, Congress passed the National Defence Education Act, which injected billions of dollars into science education:

The Congress hereby finds and declares that the security of the nation requires the fullest development of the mental resources and technical skills of its young men and women. The present emergency demands that additional and more adequate educational opportunities be made available. The defence of this nation depends upon the mastery of modern techniques developed from complex scientific principles. It depends as well upon the discovery and development of new principles, new techniques, and new knowledge. [5]

The result was a complete transformation of scientific research in America, both in the way it was carried out and how it was perceived. Belatedly, science was recognized to be an essential part of the national economy. As historian Audra J. Wolfe explains:

In 1961, Alvin Weinberg, the director of Oak Ridge National Laboratory, christened this new kind of enterprise "Big Science". The phrase seemed to capture perfectly any number of characteristics of Cold War science: the focus on large, expensive instruments; the corporate structure of scientific laboratories,; the sheer number of scientists and engineers being produced; and of course the cost. The lone scientist working at his bench had given way to the research team collaborating on massive technological machines. [6]

If politicians were slow to see the importance of science, military leaders were even slower. It may come as a surprise in today's world of drones and smart weapons, but throughout most of the 20th century—World War Two and Vietnam included—the military establishment was pathologically sceptical about any kind of new technology.

One of the key players in the fight to change this attitude was SF writer Jerry Pournelle. In 1970, he co-authored a non-fiction work called *The Strategy of Technology*, which went on to become a much-used textbook in US military academies. Right at the start, the book puts the concept of "technological warfare" into the then-current context of the Cold War:

Technological warfare is the direct and purposeful application of the national technological base . . . to attain strategic and tactical objectives. . . .. The emergence of this relatively new form of war is a direct consequence of the dynamic and rapidly advancing character of the technologies of the two superpowers and of certain of the US allies. Its most startling application to date has been the Soviet and American penetration of space and the highly sophisticated articulation of specific technical achievements in other aspects of modern conflict—psychological, political, and military. In one generation space went from the realm of science fiction to become the hallmark of superpower status. [7]

As the world came more and more to resemble the future predicted by science fiction, the status of the latter became much more respectable. Academics began to pay serious attention to the genre—and its views on the future. Of the various "futurology" studies alluded to in the earlier quote from Arthur C. Clarke, one of the most important was Alvin Toffler's *Future Shock* (1970). Unlike Clarke himself, Toffler had no connection to the SF community—he was a sociologist—but that didn't prevent him recognizing the unique importance of the genre:

> If we view it as a kind of sociology of the future ... science fiction has immense value as a mind-stretching force for the creation of the habit of anticipation. Our children should be studying Arthur C. Clarke, William Tenn, Robert Heinlein, Ray Bradbury and Robert Sheckley, not because these writers can tell them about rocketships and time machines but, more important, because they can lead young minds through an imaginative exploration of the ... issues that will confront these children as adults. Science fiction should be required reading. [8]

One example of such "required reading" has already been mentioned, back in the first chapter, "The Super-Bomb". That was Robert A. Heinlein's novel *Starship Troopers*, which is on the reading list for new recruits in the US Marine Corps. Another example features at a very different establishment, the prestigious Massachusetts Institute of Technology (MIT). This is Arthur C. Clarke's 1951 short story "Superiority"—which, in Clarke's own words, "was required reading in MIT engineering courses" [9].

"Superiority" is a cautionary tale about the dangers of adopting new technology before it's been fully tested. The action is set in a future space war, where the drive to get new weapons into the field of battle overrides long-established engineering discipline. In the case of a novel space-warping device, this leads to unforeseen but disastrous side-effects—as the defeated commander-in-chief later recounts:

> It is impossible to describe the resultant chaos. Not a single component of one ship could be expected with certainty to work aboard another. The very nuts and bolts were no longer interchangeable, and the supply position became quite impossible. Given time, we might even have overcome these difficulties, but the enemy ships were already attacking in thousands with weapons which now seemed centuries behind those that we had invented. Our magnificent fleet, crippled by our own science, fought on as best it could until it was overwhelmed and forced to surrender. [10]

As comical as Clarke's story is, even here he proved to be something of a prophet. The Vietnam War saw the United States fielding at least one piece of advanced technology that ended up being outclassed by the opposition's much more primitive equipment. The problem here wasn't so much lack of testing, but over-specialization for a role that was different from the way the technology was actually used.

The F-4 Phantom was the most advanced fighter plane of its time—and the first in history to be designed without guns. It wasn't supposed to need them, because its originally planned role involved shooting down Soviet nuclear bombers using long-range missiles. In Vietnam, however, it found itself up against small, nimble MiG fighters of an earlier generation—and embarrassingly inferior to them, as military historian David C. Isby relates:

> Anti-aircraft missiles designed for use against bombers, rather than highly agile fighters like the MiG-17, turned out to be relatively inaccurate—about 10 per cent of those fired hit their target. . ... The low accuracy of the air-to-air missile brought about a re-evaluation of the much-maligned cannon. Long considered secondary (or, in the case of the gunless F-4, superfluous) to the missile . . . the success of cannon armament in the MiG-17 saw air-to-air gunnery—long a neglected art in the USAF—being re-emphasized. [11]

The War That Never Happened

Although the Cold War wasn't a real war, with direct fighting between the two sides, it threatened to turn into one at any moment. Few people, either in the SF community or the wider population, doubted that it would be a war fought with nuclear weapons. That was a devastating prospect—but one the world was forced to live with from the 1950s right through to the 1980s. As Brian Clegg wrote in 2015:

> It would seem very difficult for young people today to comprehend, but when I was a teenager, the end of the world at our own hands was an everyday consideration. Back in the 1970s we genuinely felt that it was likely there would be a nuclear attack as an escalation of the Cold War and that life as we knew it would come to an end. [12]

The paranoid worries of the Cold War look a bit silly in retrospect—and were starting to do so as long ago as 1995, when SF critic John Clute wrote the following:

Fig. 2 The first issue of the comic book *Atomic War*, dating from November 1952, depicting a nuclear attack on America—an event that remained a real possibility throughout the Cold War (public domain image)

Now that we have survived the Cold War—which dominated world politics from 1945 to 1990—it is easy for us to look back and mock the fears of the men and women who thought that unless they remained vigilant, a world-wide nuclear holocaust was more or less inevitable. [13]

Yet the fear was real at the time, and visions of that seemingly inevitable war pervaded popular culture (see Fig. 2).

Looking beyond the populist fringes of SF, its more serious exponents were less concerned with the gory details of the war itself than its possible aftermath. One prominent author who was virtually obsessed with the subject was Philip K. Dick—whose 30-year career, from 1952 to 1982, fell entirely within the Cold War period. On one level or another it dominates most of his short stories, including several that have already been mentioned in this book: "Second Variety, "Planet for Transients", "Foster, You're Dead" and "The Hood Maker".

In addition, many of Dick's novels—from all stages of his career—are set after a nuclear war. Examples include *The World Jones Made* (1956), *Vulcan's Hammer* (1960), *The Penultimate Truth* (1964), *Dr Bloodmoney* (1965—subtitled "How We Got Along After the Bomb"), *Do Androids Dream of Electric Sheep?* (1968—later adapted, with the post-nuclear background all but deleted, as *Blade Runner*), and *Deus Irae* (co-written with Roger Zelazny, 1976).

Dick made the tacit assumption that, even if the Cold War didn't turn hot, it would continue well into the 21st century. An example of this—which has already been mentioned in a different context—is his 1967 novel *The Zap Gun*. Set in 2004, it portrays a world split into two powerful opposing factions, called Wes-Bloc and Peep-East. The former, however, is still dominated by the United States, and the latter by the Soviet Union.

Philip K. Dick wasn't the only writer to imagine the Cold War spilling over into the 21st century. In *2001: A Space Odyssey* (1968)—set, as the title suggests, at the very start of this century—Arthur C. Clarke envisaged the Americans and Russians jointly operating a space station in Earth orbit. That's accurate enough, although Clarke's station was far more ambitious than the real-world ISS. Significantly, however, it's described as having "two entrances labelled WELCOME TO THE U.S. SECTION and WELCOME TO THE SOVIET SECTION" [14]. The Soviet Union was alive and well in Arthur C. Clarke's version of 2001.

A Cold War that led to nuclear armageddon . . . a Cold War that continued into the 21st century . . . what else could SF writers think up? How about an inversion of the way things actually turned out: a peaceful Soviet victory following the political collapse of the United States? That's essentially what happens in Ben Bova's novel *Privateers*, dating from 1985. Written at the height of the controversy over Ronald Reagan's Strategic Defence Initiative, Bova's novel considers what might have happened if America had dropped the idea while the Soviets picked it up and ran with it.

The action is set in the middle of the 21st century, when the world is dominated by just one super-power: the Soviet Union. Here is how one of the novel's characters, employing the historian's present tense, explains how it came about:

> The Soviets announce that they have weapons in orbit that can shoot down ballistic missiles. The United States caves in to Soviet demands and quits

NATO.... With the Americans humbled and western Europe groping in the dark, there was no need for fighting. Paris, London, even Bonn fell all over themselves in their eagerness to make their accommodations with the new political situation. The Cold War ended almost overnight. [15]

Perhaps the most frightening possibility of all, however, is that the Cold War would simply go on and on forever. That's the situation John Brunner depicts in *The Wrong End of Time* (1971), with both sides shrunk in on themselves in an eternally defensive posture. No longer merely a Cold War, it's become, to use Brunner's own term, "the Frozen War" [16].

Isaac Asimov described a similar frozen-solid scenario in his short story "Let's Get Together", from 1957:

No one said the East, or the Reds or the Soviets or the Russians any more.... There was no hatred, even. At the beginning, it had been called a Cold War. Now it was only a game, almost a good-natured game, with unspoken rules and a kind of decency about it....

Almost at the beginning of what had been the Cold War, both sides had developed thermonuclear weapons, and war became unthinkable. Competition switched from the military to the economic and psychological and had stayed there ever since. But always there was the driving effort on each side to break the stalemate ... and that was not because either side wanted war so desperately, but because both were afraid that the other side would make the crucial discovery first. For a hundred years each side had kept the struggle even. And in the process, peace had been maintained for a hundred years. [17]

Cold War SF: Proactive or Reactive?

This book has seen numerous instances of science fiction seemingly anticipating technologies that came into use during the Cold War. Was this genuine prediction—or was it simply a reflection of developments that were inevitable, given the current trend of scientific thought? There's no simple answer to this question, and SF has had its share of both success and failure.

In *Profiles of the Future*, Arthur C. Clarke defines two "hazards of prophecy". The first of these is a failure of nerve: "even given all the facts, the would-be prophet cannot see that they point to an inescapable conclusion" [18]. SF rarely suffers from this problem—in contrast to mainstream science and engineering, which are singularly prone to failures of nerve.

The archetypal example is space travel. Throughout the 1930s, most professionals in Britain and America were glibly dismissive of the subject, despite

Fig. 3 The cover of *Galaxy Science Fiction* for September 1952, illustrating an article in the magazine by Willy Ley entitled "Space Travel by 1960?" (public domain image)

the fact that rocket theory was already well established. Even after the V-2 showed rockets to be a practical proposition, the scientific establishment continued to denigrate space travel well into the 1950s. Most notoriously, just a year before the launch of Sputnik 1, the British astronomer Sir Richard Woolley confidently asserted that "space travel is utter bilge" [19].

Four years before that announcement—and with much greater accuracy—the cover of the September 1952 issue of *Galaxy Science Fiction* speculated on the possibility of "Space Travel by 1960?" (see Fig. 3).

Writing inside the magazine, Willy Ley remarks that what was "old news" for SF fans was finally becoming a topical talking-point for everyone else:

> This issue's cover is something of "instant recognition" to science fiction readers—it shows a spaceship take-off. Science fiction readers would have recognized such a picture even 20 years ago. Now, however, the same picture

might be on the cover of any magazine and the majority of the readers of that magazine would know what it is supposed to show. That is vast progress. [20]

When it comes to predicting the future, science fiction rarely suffers from a failure of nerve. One of its most prominent failures, in fact, was the exact opposite. It was assumed that once space travel had been achieved, people would rapidly push further and further outwards—first to the Moon, then to Mars, Jupiter, Saturn and beyond. The idea that humans would reach the Moon by the end of the 1960s was common currency in mid-century SF. On the other hand, the idea that they would have got no further 50 years later—and indeed, retreated back into Earth orbit—was something SF totally failed to predict.

Arthur C. Clarke's second hazard of prophecy is the failure of imagination: "when the really vital facts are still undiscovered, and the possibility of their existence is not admitted" [21]. Again, the best-known example of this comes not from SF but the world of academic science. As we saw in the first chapter, "The Super-Bomb", people like Ernest Rutherford and Albert Einstein—who played key roles in the development of atomic theory—were utterly dismissive of suggestions that it might ever have a practical use.

Science fiction writers took a less short-sighted view, following the latest scientific developments and imagining where they might lead. While Rutherford and Einstein were scoffing at the idea of an atomic bomb, H. G. Wells—in his novel *The World Set Free* (1914)—showed how it might be employed in a future war. He got many of the technical details wrong, and the bombs he describes are nothing like their real-world counterparts. What he did get right, however, was the most important point of all: that atom bombs would be so much more devastating than conventional ones that they would change the whole structure of world politics.

Science fiction did, however, suffer a failure of imagination of its own—and that was in the area of electronics. The early SF magazines showed no shortage of imagination *per se* on the topic, but it was misdirected—with its "ray gun gothic" visions of death rays and clanking, humanoid robots. It failed to predict the silicon revolution—or anything resembling it—and the miniaturization of electronic components that led to the laptop computers and smartphones of today.

It could be argued that science fiction had another impact on the Cold War, through its use as propaganda. Several examples were given in the chapter on "Mind Games"—from the anti-nuclear message of *The Day the Earth Stood Still* to the universal brotherhood of Gene Roddenberry's *Star Trek*. It's not clear, however, that the use of SF in these cases had any fundamental significance—it was simply a popular genre at the time. If sci-fi hadn't existed, the same messages would have been conveyed through other media.

From a historical perspective, a more significant—and rather disturbing—question is the extent to which SF inspired some of the more extreme lines of Cold War research. Ronald Reagan's Strategic Defence Initiative, for example—with its beam weapons and orbiting battle stations—was even given the sci-fi nickname of *Star Wars*. Added to that was the CIA's use of telepathic remote viewing, and the US Air Force's apparent interest in antigravity research. Even in hindsight, it's difficult to disentangle how much of this work was serious research, and how much was disinformation designed to confuse the enemy.

In either case, we're left with the same question. Would such activities ever have been embarked on, if science fiction hadn't laid the groundwork?

Beyond the Cold War

When science fiction looks at the future, it usually does so by identifying current trends and extrapolating them forwards. That can be very effective in fields like science and technology, where SF is most at home, because they generally advance in a linear way. When it comes to "background" elements like politics and world events, however, that simply isn't true—and extrapolation can give entirely the wrong answer.

Throughout the Cold War, SF writers tended to envisage a future in which smaller nations would gradually be subsumed into larger ones, and wars—if they occurred at all—would be on the global scale of the 20th century's two world wars. Instead, the post-Cold War trend has been in the opposite direction—towards numerous regional conflicts, and the break-up of larger countries into small, independent ones.

Nevertheless, a few SF novels of the time did manage to look beyond the geopolitics of the Cold War. Right in the middle of it, in 1969, Brian Aldiss wrote *Barefoot in the Head*, for example. As mentioned in the "Mind Games" chapter, this deals with an attack on Europe by a Middle-Eastern country using relatively low-tech chemical weapons. It's by no means an accurate prediction of early 21st century events, but it's closer to the mark than the communist-versus-capitalist fixation of many of Aldiss's contemporaries.

It's missing the point, however, to criticize Cold War SF too harshly for being rooted in the political obsessions of its time. Most authors didn't write their novels for posterity, in order to impress people yet unborn with their prophetic powers. They wrote them for readers at the time, in the hope they would purchase as many copies as possible, as quickly as possible. So it's no surprise that it was these readers' interests, and their concerns, that SF authors pandered to.

The biggest of all those concerns was the proliferation of nuclear weapons. Soon after the start of the Cold War, the *Bulletin of the Atomic Scientists* created the concept of the "Doomsday Clock", which indicates the imminence of nuclear armageddon in terms of a symbolic number of "minutes to midnight". As the *Bulletin*'s website explains:

> The Doomsday Clock is a design that warns the public about how close we are to destroying our world with dangerous technologies of our own making. It is a metaphor, a reminder of the perils we must address if we are to survive on the planet. [22]

When it first appeared in 1947, the clock was set to a "baseline" value of seven minutes to midnight. Since then, that figure has been raised or lowered according to the seriousness of the threat relative to the situation in 1947. A high value (a large number of minutes) is a good thing, while a low value is bad. Fig. 4 shows the ups and downs of the Doomsday Clock over the 70 year period 1947–2017.

Some of the trends are easy to explain. The Doomsday Clock was set to a low value—corresponding to a precarious situation—in the 1950s, which saw both the United States and the Soviet Union testing megaton-scale H-bombs with alarming frequency. A second low period occurred in the 1980s, when the delicate stand-off was threatened by Ronald Reagan's talk of a "Star Wars" space-based weapon system. Soon afterwards, however, there was an abrupt increase in the "minutes to midnight"—a diminishing of world tension—with the collapse of the Soviet Union in 1991.

Since then, the situation has gradually worsened again. That's partly because the nuclear threat hasn't gone away completely. Thousands of warheads are still stockpiled by Russia and America, while other countries have developed nuclear aspirations of their own. At the time of writing, the biggest threat appears to come from North Korea, which not only possesses viable nuclear weapons but also long-range ballistic missiles capable of delivering them.

In spite of that, it's surprising to see that the Doomsday Clock is almost as close to midnight now as it was at the lowest point of the Cold War. The reason, however, has nothing to do with nuclear weapons. Since 2007, the *Bulletin of the Atomic Scientists* has factored in another threat: that of man-made climate change. Although slower and less dramatic than "pushing

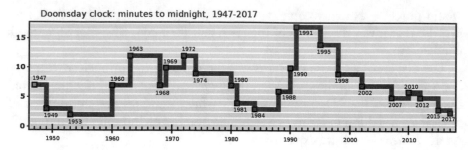

Fig. 4 History of the Doomsday Clock from 1947 to 2017. A lower numerical value (nominally "minutes to midnight") corresponds to a more precarious world situation (public domain image)

the nuclear button", the result would ultimately be just as catastrophic for life on Earth.

Throughout the Cold War—and despite all that hand-wringing about nuclear armageddon—the population of Earth was happily pumping massive amounts of carbon dioxide into the atmosphere. Although the warming effect of such emissions had been understood since the 19th century, the threat it posed to the wellbeing of the planet was hardly ever touched on—even by scientists.

Ironically, people did worry about a form of "anthropogenic climate change" during the Cold War—but, in keeping with the times, it was connected to their favourite subject of World War Three. "Nuclear winter" was mooted as a possible consequence of the widespread use of atomic weapons—and it was pretty much the exact opposite of global warming. As the atmospheric scientist Richard Turco wrote in 1983:

> Global nuclear war could have a major impact on climate—manifested by significant surface darkening over many weeks, subfreezing land temperatures persisting for up to several months, large perturbations on global circulation patterns, and dramatic changes in local weather and precipitation rates—a harsh "nuclear winter" in any season. [23]

For a wider perspective on climate catastrophes, it's necessary—as usual—to turn to science fiction. During the 1960s and 70s, in among all those nuclear doomsday scenarios, a few writers chose to focus on environmental catastrophes instead. An anthology of such stories, called *The Ruins of Earth*, was put together by the author Thomas M. Disch in 1973. In the book's introduction, he prophetically anticipates the Doomsday Clock's realization that ecological disaster is just as bad as the nuclear variety:

Nuclear catastrophe and its aftermath was … the worst nightmare we could imagine. … One learned to live with the bombs largely by looking the other way, by concentrating on the daytime, suburban side of existence. And here we are, a quarter of a century after Hiroshima, and the bombs still haven't dropped. … It is the daytime, suburban side of existence that has become our nightmare. In effect the bombs are already dropping—as more carbon monoxide pollutes the air of Roseville. [24]

It's significant that Disch singles out carbon monoxide—which, together with other industrial pollutants, has a fairly obvious impact on air quality— rather than carbon dioxide, with its subtler and more insidious effects on global temperatures. This reflects the general tenor of the "environmental" SF of that time, which was less concerned with climate change than with health issues.

There were a few exceptions, however—such as James Blish's novella "We All Die Naked" from 1969. Set in a New York of the not-too-distant future, rising sea levels have turned the streets into canals. The blame for this, in Blish's own words, lies with "the burning of fossil fuels, begun in prehistory among the peat bogs". As he goes on to explain:

Carbon dioxide is not a poisonous gas, but it is indefatigably heat-conservative, as are all the other heavy molecules that had been smoked into the air. In particular, all these gases and vapours conserved solar heat, like the roof of a greenhouse. In due course, the Arctic ice cap, which had only been a thin sheet over a small ocean … melted, followed by the Greenland cap. Now the much deeper Antarctic cap was dwindling, dumping great icebergs into the warming Antarctic Ocean. [25]

When it first appeared, "We All Die Naked" made hardly any impact—and it's rarely been reprinted since. On this issue, it seems, Blish was a prophet without honour. Readers wanted their sci-fi to warn them about things they were already aware of, like nuclear war, rather than things they'd never heard of, like climate change.

It's the other way around today. Few people spend much time worrying about apocalyptic nuclear war, but almost everyone—with any intelligence, anyway—is concerned about global warming. Science fiction is no exception, as Brian Clegg pointed out in 2015:

So common is science fiction with a climate change theme at the moment, particularly for the young adult market, that it has been given a subcategory of its own, known as cli-fi. There is plenty to work on as a result of the dire warnings of the climate scientists. Take sea-level rise. If the entire Greenland ice sheet were to melt and end up in the ocean, it would raise sea levels by 23 feet. This would have a catastrophic impact on cities near to sea level like New York. [26]

The modern world has other worries besides climate change—and one of them comes straight from the pages of science fiction. According to some people, artificial intelligence (AI) poses a serious threat to humanity's long-term existence. That's something SF writers have been hammering on about ever since Karel Čapek coined the word robot in 1920. The difference now is that many academics are saying the same thing. To quote a BBC report from 2012:

Cambridge researchers are to assess whether technology could end up destroying human civilization. The Centre for the Study of Existential Risk (CSER) will study dangers posed by biotechnology, artificial life, nanotechnology and climate change. The scientists said that to dismiss concerns of a potential robot uprising would be "dangerous". Fears that machines may take over have been central to the plot of some of the most popular science fiction films. Perhaps most famous is Skynet, a rogue computer system depicted in the *Terminator* films. Skynet gained self-awareness and fought back after first being developed by the US military. But despite being the subject of far-fetched fantasy, researchers said the concept of machines outsmarting us demanded mature attention. [27]

Two years after the formation of CSER, one of its advisors told the BBC: "the development of full artificial intelligence could spell the end of the human race". The advisor in question wasn't just anyone—it was the world-famous theoretician Stephen Hawking. He went on to explain how an AI takeover might occur:

It would take off on its own, and redesign itself at an ever increasing rate. Humans, who are limited by slow biological evolution, couldn't compete, and would be superseded. [28]

Another person who is equally worried about the AI threat is Elon Musk, the founder of Tesla Motors and SpaceX. In August 2017, he tweeted:

If you're not concerned about AI safety, you should be. Vastly more risk than North Korea. [29]

Perhaps the *Bulletin of the Atomic Scientists* should add AI to the Doomsday Clock, alongside climate change and nuclear armageddon. Musk emphasized the point with another tweet the following month:

Competition for AI superiority at national level most likely cause of World War
Three in my opinion. [30]

This statement would have come as a huge surprise to the Cold War SF
community, most of whom were convinced that World War Three—if it ever
happened—would be the result of a clash between the communist East and
capitalist West. Taking a wider view, however, that community was resound-
ingly right about the way the future would turn out. The world of the 21st
century, just as it always was in SF, is dominated and defined by its technology.

In fact it's become something of a cliché to describe any emerging technology as "like something out of science fiction". That's been said of everything
from AI and Tesla's self-driving cars to 3-D printing—which is often likened
to *Star Trek*'s "replicators"—and wearable technology. In the form of wrist-
worn gadgets, the latter has a long and venerable SF heritage. As the *TV Tropes*
website puts it, "in fiction, a bracelet is never just a bracelet and a watch never
just tells time" [31].

It would have been unthinkable half a century ago for a mainstream news
item to make reference to a work of SF—if only because the allusion would
have been lost on most readers. Now, with the rise of the blockbuster sci-fi
movie, such allusions are commonplace. Here's an example from Britain's
most prestigious newspaper, *The Times*, in October 2017:

A billboard installed at Piccadilly Circus in central London will use surveillance
technology to broadcast adverts based on people's age, gender and mood.
Cameras hidden in the screen will also detect the make, model and colour of
passing cars, allowing brands to create commercials for drivers of particular
vehicles. The system evokes *Minority Report*, a film from 2002 in which billboards scanned the retinas of passers-by to show them personalised adverts. [32]

The reference here is to a scene in Steven Spielberg's *Minority Report*, which
shows the protagonist—John Anderton, played by Tom Cruise—walking
through a shopping mall. He's recognized, and directly addressed, by one
electronic advertisement after another:

– Lexus. The road you're on, John Anderton, is the one less travelled . . ..
– John Anderton! You could use a Guinness right about now . . .
– John Anderton, member since 2037 . . .
– Get away, John Anderton, forget your troubles . . . [33]

Fig. 5   The crew of ISS Expedition 21 pay homage to *Star Trek* in this official publicity shot (NASA image)

When Oscar Wilde wrote that "life imitates art" he wasn't thinking about electronic billboards in Piccadilly Circus—but if he was alive today, he might have been. The real world is now just as likely to emulate SF as any other genre. When Gene Roddenberry created *Star Trek* at the height of the Cold War, his depiction of so many different nationalities working together on the USS Enterprise may have seemed far-fetched. Yet just this kind of cooperation is now a matter of routine on the International Space Station.

Coincidence, or cause-and-effect? Whatever the case, the link was officially acknowledged in 2009, when the crew of ISS Expedition 21—made up of two Russians, two Americans, one Belgian and one Canadian—posed for a publicity shot dressed in *Star Trek* uniforms, with a silhouette of the USS Enterprise in the background (see Fig. 5).

References

1. A.C. Clarke, *Profiles of the Future* (Gollancz, London, 1962), front cover
2. A.C. Clarke, *Profiles of the Future* (Pan Books, London, 1973), p. 9
3. L. del Rey, Behind the Sputniks, in *Fantastic Universe* (April 1958), pp. 56–65
4. M. Reynolds, Combat, in *Analog Science Fact-Fiction* (February 1961, UK edition), pp. 88–124
5. F. Kaplan, Can America Ever Have Another Sputnik Moment? *Slate* (June 2012), http://www.slate.com/articles/technology/future_tense/2012/06/sputnik_and_american_science_why_another_sputnik_moment_would_be_impossible_today_.html
6. A.J. Wolfe, *Competing with the Soviets* (Johns Hopkins University Press, Baltimore, 2013), pp. 40, 41
7. S.T. Possony, J.E. Pournelle, F.X. Kane, *The Strategy of Technology* (online version, 1997), https://www.jerrypournelle.com/sot/sot_1.htm
8. A. Toffler, *Future Shock* (Pan Books, London, 1971), p. 384
9. A.C. Clarke, *Astounding Days* (Gollancz, London, 1990), p. 99
10. A.C. Clarke, Superiority, in *Expedition to Earth* (New English Library, London, 1976), pp. 90–101
11. D.C. Isby, *Fighter Combat in the Jet Age* (Harper Collins, London, 1997), p. 77
12. B. Clegg, *Ten Billion Tomorrows* (St. Martin's Press, New York, 2015), p. 138
13. J. Clute, *Science Fiction: The Illustrated Encyclopedia* (Dorling Kindersley, London, 1995), p. 264
14. A.C. Clarke, *2001: A Space Odyssey* (Arrow Books, London, 1968), p. 55, 56
15. B. Bova, *Privateers* (Methuen, London, 1986), pp. 22, 23
16. J. Brunner, *The Wrong End of Time* (Methuen, London, 1975), p. 67
17. I. Asimov, Let's Get Together, in *The Rest of the Robots* (Panther, London, 1968), pp. 76–97
18. A.C. Clarke, *Profiles of the Future* (Pan Books, London, 1973), p. 19
19. A.C. Clarke, *Profiles of the Future* (Pan Books, London, 1973), p. 26
20. W. Ley, Space Travel by 1960? in *Galaxy Science Fiction* (September 1952), pp. 90–94
21. A.C. Clarke, *Profiles of the Future* (Pan Books, London, 1973), p. 30
22. K. Benedict, Doomsday Clockwork, in *Bulletin of the Atomic Scientists* (January 2017), https://thebulletin.org/doomsday-clockwork8052
23. B. Clegg, *Armageddon Science* (St. Martin's Press, New York, 2010), p. 102
24. T.M. Disch, *The Ruins of Earth* (Arrow, London, 1975), p. 11
25. J. Blish, We All Die Naked, in *Three for Tomorrow* (Sphere Books, London, 1972), pp. 156, 157
26. B. Clegg, *Ten Billion Tomorrows* (St. Martin's Press, New York, 2015), p. 140
27. BBC News, Risk of robot uprising wiping out human race to be studied (26 November 2012), http://www.bbc.co.uk/news/technology-20501091

28. R. Cellan-Jones, Stephen Hawking warns artificial intelligence could end mankind, in *BBC* (December 2014), http://www.bbc.co.uk/news/technology-30290540

29. S. Gibbs, Elon Musk: AI vastly more risky than North Korea, in *Guardian* (August 2017), https://www.theguardian.com/technology/2017/aug/14/elon-musk-ai-vastly-more-risky-north-korea

30. A. Hern, Elon Musk says AI could lead to third world war, in *Guardian* (September 2017), https://www.theguardian.com/technology/2017/sep/04/elon-musk-ai-third-world-war-vladimir-putin

31. TV Tropes, Gadget Watches, http://tvtropes.org/pmwiki/pmwiki.php/Main/GadgetWatches

32. M. Moore, Personalised ads delivered by the billboard that's got its eye on you, in *The Times* (October 2017), https://www.thetimes.co.uk/article/personalised-ads-delivered-by-the-billboard-that-s-got-its-eye-on-you-hrxwrrb3z

33. S. Frank, J. Cohen (screenplay), *Minority Report* (20th Century Fox, 2002)

Printed in the United States
By Bookmasters